DARK MATTER & DARK ENERGY

The Hidden 95% of the Universe

BRIAN CLEGG

D0270810

ICON

Published in the UK and USA in 2019
by Icon Books Ltd, Omnibus Business Centre,
39–41 North Road, London N7 9DP
email: info@iconbooks.com
www.iconbooks.com

Sold in the UK, Europe and Asia
by Faber & Faber Ltd, Bloomsbury House,
74–77 Great Russell Street,
London WC1B 3DA or their agents

Distributed in the UK, Europe and Asia
by Grantham Book Services,
Trent Road, Grantham NG31 7XQ

Distributed in the USA
by Publishers Group West,
1700 Fourth Street, Berkeley, CA 94710

Distributed in Australia and New Zealand
by Allen & Unwin Pty Ltd,
PO Box 8500, 83 Alexander Street,
Crows Nest, NSW 2065

Distributed in South Africa
by Jonathan Ball, Office B4, The District,
41 Sir Lowry Road, Woodstock 7925

Distributed in India by Penguin Books India,
7th Floor, Infinity Tower – C, DLF Cyber City,
Gurgaon 122002, Haryana

Distributed in Canada by Publishers Group Canada,
76 Stafford Street, Unit 300
Toronto, Ontario M6J 2S1

ISBN: 978-178578-550-4

Typeset in Iowan by Marie Doherty

Printed and bound in Great Britain
by Clays Ltd, Elcograf S.p.A.

Hot Science is a series exploring the cutting edge of science and technology. With topics from big data to rewilding, dark matter to gene editing, these are books for popular science readers who like to go that little bit deeper ...

AVAILABLE NOW AND COMING SOON:

Destination Mars:
The Story of Our Quest to Conquer the Red Planet

Big Data:
How the Information Revolution
is Transforming Our Lives

Gravitational Waves:
How Einstein's Spacetime Ripples Reveal the Secrets
of the Universe

The Graphene Revolution:
The Weird Science of the Ultrathin

CERN and the Higgs Boson:
The Global Quest for the Building Blocks of Reality

Cosmic Impact:
Understanding the Threat to Earth from Asteroids
and Comets

Artificial Intelligence:
Modern Magic or Dangerous Future?

Astrobiology:
The Search for Life Elsewhere in the Universe

Dark Matter & Dark Energy:
The Hidden 95% of the Universe

Outbreaks & Epidemics:
Battling Infection From Measles to Coronavirus

Rewilding:
The Radical New Science of Ecological Recovery

Hacking the Code of Life:
How Gene Editing Will Rewrite Our Futures

Origins of the Universe:
The Cosmic Microwave Background
and the Search for Quantum Gravity

Behavioural Economics:
Psychology, Neuroscience,
and the Human Side of Economics

Hot Science series editor: Brian Clegg

ABOUT THE AUTHOR

Brian Clegg is the author of many books, including most recently *Professor Maxwell's Duplicitous Demon* (Icon, 2019) and *The Reality Frame* (Icon, 2017). His *Dice World* and *A Brief History of Infinity* were both longlisted for the Royal Society Prize for Science Books. Brian has written for numerous publications including *The Wall Street Journal*, *Nature*, *BBC Focus*, *Physics World*, *The Times*, *The Observer*, *Good Housekeeping* and *Playboy*. He is a Fellow of the Royal Society of Arts, the editor of popularscience.co.uk and blogs at brianclegg.blogspot.com.

www.brianclegg.net

For Gillian, Rebecca and Chelsea

ACKNOWLEDGEMENTS

My thanks to the team at Icon Books involved in producing this series, notably Duncan Heath, Robert Sharman and Andrew Furlow. Particular thanks to the late Sir Patrick Moore for being my first inspiration to take an interest in astronomy and my lecturers at the University of Cambridge for making cosmology one of the most inspiring aspects of my degree course.

CONTENTS

THINGS AIN'T WHAT THEY SEEM TO BE $\qquad$ 1

Hidden depths

The universe is a big place: phenomenally big by the scale of anything we can directly experience. To be honest, we don't actually know *how* big it is, though the part we can see is around 91 billion light years across. Given that a light year (the distance light travels in a year) is around 9.46 trillion kilometres (5.9 trillion miles)*, that's a fair distance. And as the universe contains many billions of galaxies, the majority of which hold billions of stars, there is a whole lot of stuff out there. Yet in the twentieth century, two challenges to our understanding of the nature of the universe have meant that what we once thought was *everything* appears to be only around 5 per cent of reality.

---

* To give a feel for the scale of a light year, to travel a light year you would have to circumnavigate the Earth about 236,500,000 times.

Once, our picture of what made up the universe was simple. Ancient Greek philosopher Aristotle made use of an existing theory of four elements – earth, water, air and fire – and added a fifth, the quintessence or aether, which he thought made up the unchanging heavens. As astronomy and science advanced, it became clear that Aristotle's model was flawed. By the nineteenth century, it was possible to detect the chemical elements that existed in the stars – and they proved to be the same as those that were found on Earth. By the twentieth century, the five elements had been replaced by around 94 natural elements of the periodic table, each made up of a very small number of fundamental particles: protons, neutrons and electrons.

Although later in the twentieth century, those protons and neutrons would be discovered to have smaller components, this broad picture of everything being made of a handful of simple building blocks held. Yet a series of events was to shatter this simplistic picture. If science has one commandment, it's: 'Things are more complicated than we thought.' And the idea that all that existed in the universe could be made up from a few particles of matter, light, and four forces* would not stand the test of time. Gradually, oddities began to be uncovered.

Science is frequently misunderstood as being about the collection of facts. While fact-collecting certainly happens, it's not really the core of the discipline. As American biologist Stuart Firestein pointed out in his book *Ignorance*, it's not what we *know* that's important to science: 'Working scientists don't get bogged down in the factual swamp

---

* The four forces are gravity, electromagnetism, plus the strong and weak nuclear forces. The latter two hold together particles in the nucleus and are responsible for particle transmutation, respectively.

because they don't care all that much for facts. It's not that they discount or ignore them, but rather that they don't see them as an end in themselves. They don't stop at the facts; they begin there, right beyond the facts, where the facts run out.'

And the facts of what the universe was made of had begun to run out by 1933 for a Swiss astronomer named Fritz Zwicky.

Zwicky's misbehaving galaxies

Zwicky, it is generally agreed, was something of a character. Born in Varna, Bulgaria in 1898, son of an influential businessman and politician of Swiss extraction, he was sent to live with his extended family in Switzerland when he was six. He studied maths and physics at Einstein's alma mater, the Swiss Federal Polytechnic (Eidgenössische Technische Hochschule) in Zurich. Although he remained a Swiss citizen, he spent most of his working life at the California Institute of Technology, where he was based from 1925.

Like his younger counterpart, English astrophysicist Fred Hoyle, Zwicky was known for the richness of his imagination, producing many ideas in astrophysics and cosmology. Inevitably some of these concepts were little more than speculation: it went with the territory. In fact, it was common in physics circles even as late as the 1970s to comment that 'There's speculation, then there's more speculation, then there's cosmology.' But even by cosmological standards, some of Zwicky's ideas were outlandish.

Also like Hoyle, Zwicky's outstanding imagination did not stop him having impressive hits. Along with German astronomer Walter Baade, he was the first to give serious consideration to the concept of a neutron star – a star that had collapsed to become an incredibly dense collection of neutrons.* He coined the term 'supernova' for the explosion resulting in such a star forming, and discovered many supernova† remains.

Another significant contribution by Zwicky originated in Einstein's general theory of relativity. This theory describes the interaction between matter and spacetime (see page 92) – matter distorts the spacetime near it, producing the effects we describe as gravity. Inherent in general relativity is the idea that massive objects cause rays of light to bend, as the space the light passes through is warped by the matter. As American physicist John Wheeler put it, 'Spacetime tells matter how to move; matter tells spacetime how to curve.' Zwicky realised that this effect was similar to that produced by an ancient optical device – the lens.

Lenses (given the Latin name of a lentil because they are similarly shaped) bend the path of light by different amounts, depending on the thickness of the glass the light hits. The circular shape modifies the light's path by an increasing amount as we get further from the centre, because the glass is at a more extreme angle to the light, meaning that the lens collects together rays of light hitting it at various points and focuses them.

---

* Neutrally charged particles found in the nucleus of atoms.
† The supernova features frequently in this book, so I'd like to get one bone of contention out of the way. As 'nova' is Latin for 'new', and the Latin plural of nova is novae, the plural of supernova is usually given as supernovae. However, supernova is not a Latin word (if it had been, it would have been an adjective, not a noun), and as such I prefer and use throughout the plural 'supernovas'.

Thinking about the way a lens worked, Zwicky realised that an extremely massive object such as a galaxy could have a similar effect on passing light. If we imagine light coming from a distant object behind a galaxy, some of the light would attempt to pass around the edge of the galaxy. But the huge mass of the galaxy would bend the light beams inwards from all sides, focusing the light a great distance ahead of the galaxy. If we were positioned appropriately, and the image was cast in such a way that it wasn't washed out by the light from the galaxy, this 'gravitational lensing' would mean that we could see a very distant object by using the intervening galaxy as if it were the lens in a vast telescope.

Gravitational lensing involves something we can see – a galaxy – having a gravitational effect on passing light. But Zwicky's greatest discovery would involve a gravitational effect that appeared to come from an invisible source. He had been studying a collection of galaxies known as the Coma

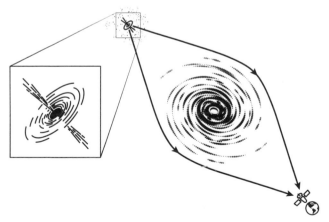

Light from a distant object is focused by an intervening galaxy acting as a gravitational lens.

Adapted from an image released by ESA/ATG medialab

Cluster. Galaxies are vast bodies – our own Milky Way, for example, a fairly average large galaxy, is over 150,000 light years across. Containing billions of stars each, galaxies have a huge gravitational influence on their surroundings and as a result readily form clusters with other galaxies, held together by gravity.

The Coma Cluster is located about 320 million light years away from us and contains over 1,000 galaxies – as the nearest neighbouring cluster to our local cluster, the one occupied by the Milky Way (the Virgo Supercluster), it has inevitably been of great interest to astronomers. Yet when Zwicky started to analyse the behaviour of the cluster in 1933, he found something odd. It should not have held together.

On the whole, things in the universe spin around. We're familiar with this being the case in our own solar system. The Earth rotates on its axis once a day and orbits the (rotating) Sun once a year, as do the other planets, each with their own distinct period. Planets, moons, stars, solar systems, galaxies, galactic clusters all spin around. This is a result of the way that they formed. These structures were produced from clouds of gas and dust, pulled together by the force of gravity. If those clouds were perfectly symmetrically dispersed through space, then they could collapse without developing a spin. But in reality, it is far more likely that there will be more matter on one side than the other. As the matter is attracted inwards, the result of this imbalance is that the whole collection of stuff begins to rotate.

It's no surprise, then, that the Coma Cluster rotates. Zwicky combined the speed of the cluster's rotation with an approximation of the amount of matter in the cluster – and got a shock. It seemed that the cluster was spinning so quickly that it should fly apart, like a poorly placed chunk of clay on a fast-moving potter's wheel. Gravity can only keep bodies in orbit at the right speed. If an orbiting body travels too fast, it will exceed the 'escape velocity' of the system and fly away. And according to Zwicky's calculations, the Coma Cluster was rotating not just a little too fast but many times too quickly.

Zwicky estimated that the cluster should have needed 400 times more mass to remain stable. (Since Zwicky's time, this figure has been reduced, but the cluster still rotates far too quickly for the assumed amount of matter present.) He decided that this could only be caused by large amounts of matter in the cluster that could not be detected. He called this unknown material *dunkle Materie* in German, which translated as 'dark matter'.

It might seem odd that such an important result was largely ignored at the time. However, Zwicky's reputation for inventiveness had the downside that, while his ideas were usually noted, they weren't always taken further. It was probably assumed that the effect was considerably smaller than Zwicky had calculated. Bear in mind that it required a calculation of the amount of matter in a distant collection of at least a thousand galaxies, each of which contained vast numbers of stars. There was a lot of approximation (scientific language for educated guesswork) going on.

It's also the case that Zwicky's idea of dark matter did not sound as exciting as it does today. Any dark matter was just that – perfectly ordinary matter that happened to be dark. It was assumed to be a combination of dust, low-output stars, planets, and more that had not been considered by making use of the observable, light-emitting matter. This wasn't even a new concept – Scottish physicist William

Thomson, Lord Kelvin, had made similar if less dramatic observations on the rotation of the Milky Way in 1904,* showing that a considerable amount of the matter in the galaxy was dark, as did other astronomers in the intervening period, particularly the Dutch astronomer Jan Oort in 1932.

Later, though, it would be realised that ordinary matter that did not emit light – even with the addition of the exotic concept of black holes – would simply not provide enough mass to account for this odd behaviour. There was something new and different out there. Far more of it than there was ordinary matter. Dark matter had arrived.

The expansion dilemma

By the 1990s, a second shock echoed through the small world of astrophysicists and cosmologists. It was the culmination of a breakthrough made in 1929. Then, American astronomer Edwin Hubble published data on the red shift of galaxies. We'll come back to red shift later on, but this is a means to identify the velocity of a light-emitting object. Hubble's data showed that with a few local exceptions, all galaxies were heading away from our own Milky Way. And the further a galaxy was away, the greater its red shift – the faster it was going. When plotted on a graph, this relationship roughly grew in a straight line, an observation that would be given the name 'Hubble's law'. This despite Hubble himself never

---

* Even the term 'dark matter' had already been used by French mathematician Henri Poincaré when referring to Lord Kelvin's calculation. Poincaré referred to Kelvin's missing material as *matière obscure*.

doing much with the interpretation of his data, being happy simply to collect it.

The data was used to justify the idea that the universe was expanding, a picture we now accept as a commonplace. But there was one thing that wasn't known: how rapidly that expansion was slowing down. That the rate of expansion should be slowing seemed inevitable, due to the influence of gravity. According to general relativity, the expansion should be countered by the gravitational effects of all the matter in the universe. It seemed unavoidable that there would be a gradual slowing of the expansion of space.

There were two possible outcomes of this braking effect. If the expansion of space was not fast enough, it would eventually be overcome, and space would start to contract, leading to a massive collision nicknamed the big crunch (the opposite of the big bang). If the expansion was too fast for gravity to totally overcome, the rate at which the galaxies moved away from each other would slow down, but would never reverse, leading to a universe that forever thinned out.

Until the 1990s, there was no good way to discover how far away distant galaxies were, to pair up with the red shift information on how fast they were moving. But by then, new techniques based on an understanding of the behaviour of a particular type of supernova were making it possible to combine distance with red shift and get a better picture of how the rate of expansion was slowing. In 1997, two teams raced to achieve sufficient data to quantify this.

Both would reach a conclusion at around the same time, and the outcome was a huge shock. All the evidence was that the rate at which the universe was expanding was not falling with time, but rather it was growing. Something unknown was adding energy to drive the expansion of space, accelerating the rate at which galaxies separated from each other. Without any idea what could be causing this, astrophysicists, taking the term from American cosmologist Michael Turner, termed the phenomenon dark energy. The name tells us nothing about what is involved. It might just as well have been called factor X or unizap.

As more data came in, it became possible to estimate just how much energy was required to cause this acceleration. Locally, the effect is tiny. It requires less than a joule of energy* for every cubic metre of space to provide such acceleration. But add that up across the whole universe and it is the equivalent of a vast amount of energy. Thanks to Einstein's familiar equation $E=mc^2$ (see page 26), we can equate energy and mass of matter. If we convert the estimate for the required amount of dark energy into mass, there is around fourteen times as much mass/energy in dark energy as there is in all the familiar visible matter in the universe, or around twice as much as there is in ordinary matter and dark matter combined.

Darkness prevails

If the theories of dark matter and dark energy are correct, around 27 per cent of the universe is dark matter and 68 per cent dark energy, leaving only around 5 per cent as everything that we directly observe. This is a big issue. Yet the nature of these phenomena is still under debate. Dark matter may not even exist. The scale of dark energy comes up

---

* Less than the amount of energy required to keep a typical low-power LED light bulb going for ⅕ of a second.

differently depending on the way that it is measured and is totally at odds with what is otherwise our most accurate physical theory, quantum mechanics. Arguably, this is the most exciting aspect of modern science.

Before we can understand the science behind the study of dark matter and dark energy, we need to fill in some fundamentals about the universe, what makes it up and how it operates. And what better place to start than with one of the oldest disciplines of science – our quest to find out more about the universe without ever leaving the Earth.

EXPLORING THE UNIVERSE

2

Forget the spaceship

I am a big fan of space exploration – I believe that it is something we need to do as humans, both to satisfy our pioneering spirit and to give ourselves an escape route should the Earth ever become uninhabitable. But to understand the universe that we live in, traditional exploration will never be a practical approach. This book opened with a statement about the size of the universe – but that scale is difficult to grasp. Let's think for a moment of planning a visit to the nearest star after the Sun, called Proxima Centauri.

That star is around four light years away. This is a tiny distance compared to the Milky Way galaxy's 150,000-light-year diameter. But it's still ridiculously far from a human perspective. The fastest a human being has ever travelled with respect to the Earth was on Apollo 10 at 39,896 kilometres per hour. It sounds fast, but that's only 0.000037 times the speed of light. At that speed it would take over 100,000 years to reach Proxima Centauri.

In some ways, then, it may seem naive to say that the exploration of the universe began on 4 October 1957, when the USSR launched the first artificial satellite, Sputnik 1. Less than 60 centimetres (2 feet) across, this fragile metallic ball sprouting two double antennae was humanity's first true venture into space. Although primarily launched as a political gesture, the satellite did provide a small amount of scientific data. Sputnik's 83 kg (183 lb) mass – 51 kg of that was its batteries – made waves totally out of proportion to its capabilities. Over the next 50 years we would see probes reaching the Moon, Mars and the outer solar system, a manned expedition to the Moon and a succession of manned space stations in orbit.

However, despite never venturing out of our near neighbourhood, we shouldn't underestimate the value of satellites to improving our understanding of the universe. Some of the most impressive space explorers have been direct successors to Sputnik – unmanned satellites carrying instruments that have expanded our knowledge immensely. Innovations such as the Hubble Space Telescope and the COBE and WMAP satellites have been the true explorers of the universe on our behalf. And in doing so, they carry on a tradition of visual exploration that goes back far further than any space flight – back past Galileo's telescope and Ptolemy's surveys of the sky to the naked eye observations of the earliest humans.

When it comes to exploring the universe, we can forget spaceships other than those that launch satellites. Light is our primary vehicle of choice. People have explored the universe this way ever since they looked up and wondered at the beauty of the stars. With the naked eye it is possible to see the galaxy M31 in the constellation Andromeda. If the sky is dark enough, the Andromeda galaxy appears as a faint smudgy spot in the sky, on the side of the constellation nearest the W of the adjacent constellation Cassiopeia. Telescopes show that this smudge is, in reality, a massive spiral galaxy, but even the unaided eye enabled early explorers of the skies to see it across 2.5 million light years of space. Compare this with the furthest distance man has travelled at the time of writing – 375,000 kilometres to the Moon. Forget light years – that is 1.25 light seconds.

What's more, unless we come up with the kind of warp technology familiar from *Star Trek*, light (or equivalents such as gravitational waves and neutrinos, which move at or near the speed of light) will remain our principal means of exploring the universe. Light is the fastest thing in existence. If we could travel at half of light speed (something inconceivable with current technology), it would still take 5 *million* years to reach the Andromeda galaxy. But we can see Andromeda because light has already made the journey to us – we don't need any travel time. We will never explore most of the universe in person, but light allows us to see across immense distances.

We have come a long way since Galileo first used his telescope to discover astronomical bodies that are not visible to the naked eye. We now use the whole spectrum of electromagnetic radiation – radio, microwaves, infra-red, X-rays and more – of which visible light is just a tiny segment. And with these remarkable engines of visual exploration we can venture out to experience the strange cast of characters that populate the universe – black holes and dark matter, supernovas and quasars. This is exploration like no other.

First imaginings

Cosmology, as we've seen, is the science of the universe as a unified object, combined with the study of the laws that govern that whole. This definition of cosmology assumes, of course, that we know what 'the universe' is. The original Latin from which the word universe is derived means 'one turn', which doesn't help much. In practice, though, what we are dealing with is clear. The universe is all that physically exists, taking in everything from the smallest particle all the way up to the biggest galaxy. It encompasses all matter, all energy, collected together as a whole as if this assemblage were an entity in its own right. That's an impressive concept, and it is natural to ask questions about it.

From the earliest times, creation myths have been written to explain where that 'everything' came from. Humans are born storytellers, and creation myths are storytelling, not science. It's important to understand, though, that by calling these stories 'myths' we aren't insulting them, or those who consider them to be sacred. A myth is a story with a point. It is a mechanism to give information about a deep question, like 'Why are we here?' or 'Where did everything come from?'. A myth is not history – it is a way to help understand today's reality, through a story linking us to the past.

For the authors of the early creation myths, the universe was the Earth and the heavens. Land, sea and sky accounted for all space. Yes, there were a few oddities up there like the Sun, Moon and stars – but these were the inhabitants of sky, just as animals and people were the inhabitants of land. To a modern understanding, the logic of creation myths can seem confusing. But the vast majority of early creation myths have one thing in common – a creator. The answer to one of the biggest questions about the universe, how it came into being, was almost universally that someone made it.

This answer was neither irrational nor stupid. Thousands of years after these myths were formulated, a Victorian clergyman called William Paley used the same argument to explain how living creatures were formed. If you came across a watch lying on a beach, Paley said, you would not think it had naturally and randomly occurred. It was too complicated and too functional in its structure. Instead, you would assume that a watchmaker created it. Similarly, faced with the complex vastness of the Earth and heavens, the obvious response was: 'It could only be like that if someone made it.'

Like every other civilisation that predated them, the ancient Greeks had their creation myths. But they were the first, in the story of our understanding of the universe, to go further. It has been suggested that the ancient Greek approach might have been a reflection of their loose federation of city states with no central authority, leading to a more questioning philosophy than in a society with a rigid hierarchy. Rather than be satisfied by saying that the universe works because the gods make it work, from the sixth century BC, the ancient Greeks began to look for practical mechanisms that such creators could have employed.

A philosopher's universe

The first of the true Greek philosophers is generally held to be Thales of Miletus, born around 624 BC, who advocated looking for natural, rather than supernatural, causes for what was observed. Probably the first 'scientific' cosmology – a self-consistent picture of the universe and its origins that was built on physical forces and structures – came from one of Thales' pupils, Anaximander. Born in Miletus in Anatolia (now part of Turkey) in the first half of the sixth century BC, Anaximander did not challenge the existence of the gods; his view of the universe, however, was based on simple observation.

Unlike many creation myths that portrayed the universe emerging from water, Anaximander preferred a beginning where the universe arose from chaos in a sea of fire. This had one significant advantage – it allowed him to give a justification for a familiar natural phenomenon. Anaximander wanted to explain the lights in the sky – the Sun, Moon and stars. He reckoned that the primeval sea of fire was still out there, but the universe was protected from the flames by a huge shell (strangely, this was cylindrical rather than spherical). The shell had holes in it, and through these holes firelight escaped to provide the glow of the heavenly bodies and the heat of the Sun.

Anaximander and his contemporaries didn't give the universe much of a structure, though – after much debate over the years, this was formalised by the most famous of the Greek philosophers, Aristotle. Aristotle's model of the universe, put together in the fourth century BC at Plato's Academy, became so rigidly accepted that, with some tweaks, it remained in use for around 2,000 years. Bear this in mind if you find Aristotle's cosmology unlikely. No other model of the universe has been consistently supported for so long. It might have been wrong, but it had a kind of magnificent logic.

Aristotle put the Earth firmly at the centre of the universe, unmoving. This wasn't just a matter of egotism. It was self-evident that the Earth wasn't moving – or, surely, we would feel it. And Aristotle's idea of how everything moved, from a dropped stone to a rising pillar of smoke, depended on some elements being pulled towards the centre of the universe by a force called gravity, and others flying away from the centre of the universe thanks to another force, levity. If the Earth were not at the centre of everything, then heavy objects would fly off to some point in the sky. They wouldn't stick to the Earth.

Around the Earth in Aristotle's model were a series of invisible crystal spheres, each nested within the other. In the first sphere was suspended the Moon, then Venus, Mercury, the Sun, Mars, Jupiter and Saturn. Finally came the sphere supporting the fixed stars. This didn't mean that the stars were fixed in place – their sphere rotated – but that they all moved together, while the planets (the name comes from the Greek for 'wandering stars') moved separately against the outer sphere.

Gods still had a role in this picture. Each sphere drove the sphere within it, but something had to drive the outer sphere, the sphere of the stars, and hence powered the whole of the universe. This was the role of the 'prime mover', a deity. However, Aristotle's was still effectively a scientific cosmology – although a god was required to keep things moving, within the bounds of the universe everything functioned as a result of a heavenly clockwork.

In Aristotle's model, the light from the stars could have come from outside the universe, but this wasn't how he saw things working. According to Aristotle, the Sun was the source of all light. Everything else – the Moon, the planets and even the stars – were lit by reflected sunlight. When it was pointed out that you would expect the stars to be eclipsed, just as the Moon is eclipsed when the Earth gets between it and the Sun, Aristotle argued that the Earth's shadow did not stretch beyond Mercury, so would not eclipse the stars.

This universe seems very small to modern eyes. It was little more than a rearranged solar system with the stars tacked on the outside. Yet it was still a massive place compared with Greece or any other part of the Earth, as another Greek philosopher, Archimedes, would discover when he set out to work out the size of the whole universe. This was the first serious attempt to calculate a value that would become an essential requirement to estimate the scale of dark energy.

Archimedes was not engaged in worthless speculation – he had a serious intent in mind. Born around 100 years later than Aristotle, he was a much more practical philosopher. He indulged in sophisticated mathematics, coming close to inventing aspects of calculus, and he designed a wide range of mechanical devices, from a screw to lift water out of the ground to giant curved mirrors that would have formed the first death ray had they ever been built, focusing the heat of the Sun on a wooden ship to set it on fire.

In a little book called *The Sand-Reckoner*, Archimedes worked out how many grains of sand it would take to fill the universe. Apart from being an entertaining exercise, his purpose seems to have been to illustrate how to extend the number system. Greek maths was limited because the biggest number they had was a myriad – 10,000. If you wanted to go really large you could have a myriad myriads (100 million), but that was it. Archimedes devised a number system that started from 100 million and built up to immense scales.

To work out how many grains of sand it would take to fill the universe, he had first to decide how big the universe was – the interesting part as far as we are concerned. Using a number of basic assumptions – like the Earth is bigger than the Moon, and the Sun is bigger than the Earth – and a little geometry, Archimedes worked out that the universe was around 10 billion stades across. This is a measure based on the size of a stadium's running track. Unfortunately, it's hard to be certain what distance it represented. Stades were 600 feet, but the definition of the foot varied from city to city. However, each stadion (singular of stades) averaged around 180 metres, so his universe was around 1,800 million kilometres (1,120 million miles) across.

We now know that 1,800 million kilometres is about the size of the orbit of Saturn, which isn't a bad estimate at all of the size of the solar system, given the uncertainty of the measurement and bearing in mind that, as we saw above, the ancient Greek universe was effectively the solar system. In a tantalising extra, Archimedes pointed out that the astronomer Aristarchus had brought out a book featuring the radical suggestion that the Earth moved around the Sun, rather than the Sun around the Earth. Unfortunately, the book has been lost, so this is the only known reference to the idea.

Because having the Sun at the centre changed his geometry, Archimedes reckoned that this version of the solar system would be about 10,000 times bigger, taking the diameter up to around 18 trillion kilometres (11 trillion miles), which would contain the main planets of the solar system. The idea Aristarchus had of putting the Sun at the centre of things seems to have been largely forgotten. It was Aristotle's model that would continue to be accepted until the sixteenth century – but relatively soon, the simple idea of everything turning in perfect spheres had to be modified to match astronomical observations. Some of the planets were misbehaving.

If you plot the path that Mars takes through the sky, based on Aristotle's picture, you would expect it to follow a continuous path, making a circle around the Earth as its crystal sphere rotates. The reality is very different. Mars reverses its route in a process known as retrograde motion. It performs a slow loop-the-loop in the sky, which didn't fit with an unchangeable crystal sphere. We now know that this apparent motion is because Mars and the Earth are both travelling around the Sun, each moving at different speeds on orbits that aren't concentric circles. As a result, when seen from Earth, the orbit of Mars loops back on itself as Earth overtakes it – but this couldn't happen in Aristotle's model.

To explain this strange motion, it was suggested that a planet like Mars, instead of simply rotating around the Earth, also travelled in a separate circle called an epicycle. This was as if the sphere that held Mars had another, smaller sphere embedded in its surface, and this smaller sphere rotated too. So, Mars would travel in circles around the little sphere as that little sphere moved around the Earth with the big sphere – producing the looping motion that was observed. Even this didn't quite match observation, so instead of moving around the centre of the Earth, the large spheres orbited a point a little away from it, charmingly known as the eccentric.

This description of the universe held right through to Galileo's time. Galileo Galilei, born in 1564, wasn't the first in modern times to put the Sun at the centre of the universe. The Polish astronomer Nicolaus Copernicus had experimented with solar-centric models to explain away the complexity of planetary motion decades before. Galileo's German contemporary, Johannes Kepler, took Copernicus' idea further. He realised that he could model the behaviour of the planets better if they travelled around ellipses, squashed circles, rather than the perfect circles that the Greeks (and Copernicus) used. He also found that if the planets moved in such a way that a line joining the planet to the Sun sweeps out the same area in equal times – so it travels faster when it is nearer the Sun in its elliptical path – he could match the best observations of the timings of the movement of the planets, made by the Danish astronomer Tycho Brahe.

Galileo, famously put on trial for promoting Copernicus' system, added some logic to support the idea that the Sun was at the centre of the universe, not the Earth. As we have seen, the ancient Greek model was founded on the idea that everything rotated around the Earth. Galileo made an early telescope, and with it studied the heavens. He discovered that there were four moons orbiting Jupiter.* Here was direct evidence that not everything rotated around the Earth. Galileo's punishment for defying the religious authorities could not suppress the model that put the Sun at the centre of things. Throwing away those complex, messy epicycles just made too much sense. From the seventeenth century onwards, the familiar structure of the solar system was starting to be accepted.

The stars were no longer thought to be on a crystal sphere, although this brought a new question. How did they manage to stay up there? And if the planets were just hanging in space, what kept them rotating around the Sun? Isaac Newton would put this down to gravity, a strange force that somehow acted at a distance to keep the planets (and us)

---

* In practice there are far more moons of Jupiter – at least 79 – but Galileo discovered the most obvious four.

in place – though it would take Albert Einstein to come up with an explanation of how gravity works.

While detailed explanations would not arrive until the twentieth century, a new picture was emerging. The universe was opening up. Out there were planets, stars and more. But what are they, and where do they come from? To understand why dark matter is so significant, it's essential to know what we understood the universe to be made of before the effects of dark matter were first detected.

Build your own solar system

The universe is scattered with matter, mostly the gases hydrogen and helium, and dust. Imagine these clouds of matter floating around in space. There is no weather. There is not a touch of wind to move the matter around. But there is gravity, and though the force between the atoms of gas and particles of dust is absolutely tiny, every bit of matter is attracted by every other one. Those that are relatively near will very gradually, over aeons of time, begin to move towards each other. Initially there will only be minute amounts of matter present in any particular part of space. But with a vast amount of time, those fragments will begin to collect.

Once some matter has clumped together it will have a bigger gravitational pull and will drag in more of the gases. If there is enough matter, it will start to produce a decidedly heavy object. All that matter is pressing in on itself. As more and more particles of matter crash into the object, their energy of movement, produced by gravity, becomes heat. (Just think of rubbing your hands together – the kinetic energy of movement gets changed into heat energy by friction.) With painful slowness, the ever-growing ball will begin to heat up.

After several million years of heating up, a critical point will be reached. At this stage, three things are combining to make a remarkable reaction happen. The most common constituent of the ball, just as it is the most common substance in space, will be the simplest of the elements, hydrogen. The hydrogen atoms (or more accurately the hydrogen ions, which are hydrogen atoms with the electrons stripped off by the heat) will be pushed together under high pressure due to the gravitational attraction of a body that will by now contain many billions of tonnes of matter. The temperature at the core of this ever-growing body will have soared. And something else quite remarkable will be happening.

Particles like these hydrogen ions don't obey the rules we expect of matter on the scale of a person or a house. They are quantum particles, and rather than following ordinary mechanics, they obey quantum mechanics, the rules of how such particles behave that were discovered in the first half of the twentieth century.* One of the special characteristics of quantum particles is that, until they interact with something,

---

* Quantum physics lies behind the behaviour of the extremely small particles that make everything up. They behave quite differently from the familiar objects around us. The properties of quantum particles, such as position or momentum, have a range of values, rather than a fixed quantity, until they interact with another particle. The particle is said to be in a 'superposition' of quantum states, meaning that all that exists is the probability of the property having each of its possible values. This behaviour emerges from properties being 'quantised' – coming in chunks, rather than being selected from a continuous range. For more details, see the author's *The Quantum Age* (Icon Books, 2014).

their exact location is unsure. This uncertainty of location means that quantum particles can jump from one place to another without passing through the space in between, a process known as quantum tunnelling.

The positively charged hydrogen ions repel each other because of the electromagnetic force. Even under the temperatures and pressures that have built up, they can't get close enough to interact. But by tunnelling, a small percentage of those ions will jump to be too close to another ion. When they get so close, the strong nuclear force, which only operates over extremely short distances, takes over, attracting them together more strongly than the electromagnetic force repels them. Get past that limit and (in a slightly convoluted multi-stage process) the hydrogen ions fuse to make a new type of ion – a helium ion, the next element up in the periodic table.

In this process, a small amount of mass gets converted to energy. The equation that tells us how much energy we get when mass changes into energy is probably the best-known in all of history. It is $E=mc^2$, where E is the energy, m the mass, and c the speed of light (so, energy = mass multiplied by the speed of light squared). The speed of light is very big – so even a tiny amount of mass produces a vast amount of energy. If you could take one kilogram of matter and convert it into energy, you would get the amount of energy that a typical power station produces in six years, all in an instant. Once the fusion process begins, there is a vast outpouring of energy. This blasts out in the form of electromagnetic radiation – light. A star has formed.

Stars are the dominant building blocks of the universe. Take a look at the sky at night. Apart from the Moon and a few planets, all you will see is stars. With powerful enough telescopes we can make out billions of them. Unlike planets and the Moon, which shine with reflected light, stars are nature's lamps. And they are responsible for much more than that: they are also factories producing heavier and heavier elements over time. Some of the stars, in their old age, explode, adding heavier dust to the gases between the stars.

If the only things that formed from clouds of matter were stars, the universe would still be a dramatic place, but there would be no one around to see it. No form of life as we understand it can exist on or in a star. But not all the matter in the vicinity of a star will be drawn into this vast nuclear furnace in space. The vast majority will – the Sun, for example, contains well over 99 per cent of the matter in the solar system – but that still leaves a considerable amount of matter left surrounding the star.

Over time, it might be expected that this matter would also fall into the star, but as we have seen, like everything in the universe, as a solar system begins to form, the material in it spins around. The material around the star ends up like pizza dough, spun between the hands – it forms a disc around the star called an accretion disc. Within this disc, similar processes to star formation happen. Particles are attracted together and make bigger and bigger forms, eventually making up planets. (In principle another star could form, and this is often the case, resulting in a binary system where two stars orbit each other.)

The parts of the disc around a second-generation or later star will tend to produce rocky planets, like the Earth. Where there is less heavy matter, the composition of the planet will be more like the Sun, primarily gaseous, but without enough matter to reach a big enough mass to start fusion. The result will be a planet largely made up of gas, like Jupiter or Saturn. Although planets won't be heated up anywhere near as much as a star, they will be warmed by the same process as new particles zap into them, usually getting hot enough to produce a molten core, which may be kept in molten form (as is the case with the Earth) if there are enough radioactive elements in the planet to keep the heat flowing.

These, then, are the basic building blocks of a universe on the middle scale – planets and stars. For a long time, this was thought to be as big as universal building blocks get, but since the eighteenth century there have been suspicions about some fuzzy patches in the sky, which would eventually become known as galaxies. If it weren't for the same spinning effect that keeps the matter around a star from falling into it, there is no reason why all the stars in a galaxy would not gradually be pulled together into a single incredibly large lump by gravitational attraction. But like stars, galaxies also are spinning, and this keeps them in a disc-like structure, often spiral, like a swirling liquid heading down a plughole.

There are also collections of galaxies – clusters and superclusters – making up larger structures across the universe, such as the Coma Cluster that alerted Zwicky to the possible existence of dark matter. But the addition of the galaxy has given us our basic building blocks to construct a universe. Stars at the heart of everything, planets formed around stars, and galaxies made up of collections of stars. Of course, there are plenty of other inhabitants in the universal zoo – and we will come across them later – but we have now met what we need to start thinking about matter and putting into context Fritz Zwicky's strange discovery of the dark variety.

The Phantom Galaxy, originally catalogued as Messier 74, a dramatic spiral galaxy around 32 million light years away.
NASA, ESA, Hubble Heritage (STScI/AURA)-ESA/Hubble Collaboration

THE MATTER OF MISSING MATTER

3

The nature of stuff

The universe may be a phenomenally big place, but taking the scientific view, there appears to be an almost child-like simplicity to the components that make the whole thing up. It has been estimated that there are about 10^{80} atoms in the bit of the universe we can see. That's 1 with 80 zeroes after it – a satisfyingly large number.* Yet pretty much all that matter is made up of just 94 elements. (Chemistry fans can probably identify twenty or so more, but they are irrelevant here, as they don't occur naturally.)

We've even got a reasonable idea of how those elements came into being. All the hydrogen in the universe, and a small amount of the other very light elements such as helium

---

* The calculation is based on the estimate that there are around 100 billion galaxies, averaging a trillion stars. (In reality 100 billion is a lowest figure – it could be significantly more.) This is then multiplied by the only reasonably accurate guess in the calculation, the number of atoms in the Sun at around 10^{57}.

and lithium, is thought to have been produced after the big bang, when the initial tiny, hyper-hot universe expanded and cooled enough for matter to form. The heavier elements up to iron were forged in stars. The power source of stars, nuclear fusion, joins lighter elements together to produce heavier ones, giving off some energy in the process, as we have seen.

This leaves a bit of an embarrassing lack of the elements heavier than iron (which is only number 26). The vast majority of the rest is assembled when a supernova occurs: these vast stellar explosions provide enough energy to overcome the serious barriers to forming heavier atoms. Some heavy atoms also appear to be produced from the dramatic collision of ultra-dense neutron stars. It used to be thought that uranium, element 92, was the heaviest naturally occurring element, but some plutonium (number 94) has been detected in space.

We can even take that simplification to another level. All of those atoms, whatever the element, are made up of just four component parts: electrons, 'up' quarks, 'down' quarks and gluons. The heavy central nucleus of each atom contains protons and neutrons, which themselves are composed of these two types of quarks and the particles responsible for holding them together, gluons. And around the outside of the atom, in a fuzz of probability, are one or more electrons.

Add in the photon, the particle that makes up light in all its glory, from the low-energy radio waves, through visible light up to high-energy X-rays and gamma rays, and we've got most of what's necessary to make up the stuff of the universe.

If you look up the standard model of particle physics (see Appendix, page 149) you'll find a good number of other component particles. There are four more quarks; two electron-like particles, the muon and the tau particle; three types of neutrino; the Z and W bosons; and the twenty-first century's favourite particle, the Higgs boson. All of these are necessary to make reality work. They have all been detected. But these are the background actors of the world of stuff. We don't need to give them too much consideration when thinking about how the matter of the universe is put together.

What the unexpected rotational behaviour of galaxies seems to require, though, is something more – something that in all likelihood goes outside the standard model. Dark matter.

The name is perverse, getting reality almost back to front. This stuff isn't dark at all. A substance that is dark absorbs much of the light that falls on it because of an electromagnetic interaction between the photons of light and the electrons in its atoms. Think, for example, of the ultimate example of darkness, a black hole.* It is black because any light foolish enough to get closer than its event horizon will never escape. In a sense, dark matter is the antithesis of this – it absorbs no light whatsoever. It doesn't interact with light electromagnetically at all, though it does have a gravitational effect on it. In reality, dark matter is transparent matter.

Electromagnetically speaking it's as if dark matter doesn't exist – yet gravitationally, something is producing the effect that is ascribed to it. As we have seen, that effect was first commented on by Fritz Zwicky, who was promptly ignored.

---

* In practice, black holes are not strictly dark because of the radiation given off by matter accelerating as it plunges in towards the collapsed star, but I'm referring here to the black hole itself, from which no light can escape, making it the ultimate in darkness.

Slow realisation

It was not a year or two, but a good forty years after Zwicky first put forward his hypothesis that the concept of dark matter was taken any further. In the early 1970s, similar observations to Zwicky's had been made about small satellite galaxies orbiting larger ones. But the real change in attitude was the result of work from one of two great female astronomers from the 1960s and 70s who, shockingly now, may have missed out on a Nobel Prize because of their gender.

The earlier failed recipient was British radio astronomer Jocelyn Bell (now Bell Burnell) who discovered the fast-rotating neutron stars known as pulsars. Particularly outrageously in her case, Bell made the discovery, but the Nobel Prize was awarded to her thesis advisor, Antony Hewish. (Fred Hoyle, working in the same department at Cambridge, kicked up a huge fuss about this, but to no avail.) However, the champion of dark matter who was also passed over for the prize, was American astronomer Vera Rubin.

Shortly after joining the Carnegie Institute in 1965, Rubin started work with instrument maker and astronomer Kent Ford and it was with Ford that she would make her breakthrough observations leading to the revival of dark matter. Ford had produced a device that amplified visual input electronically, making it possible to make spectroscopic readings (of which more soon) of a galaxy several times in one night, where before it would have taken tens of hours. For the few years previous to this, Rubin had been studying the rotation of galaxies, such as our nearest large neighbour the Andromeda galaxy, and discovered something very strange.

When a solid disc like a CD is rotating, a point on the disc that is nearer the edge travels faster than a point near the centre. This is inevitable as the further out a part of the disc is, the greater distance that point has to travel in any particular elapsed time. With a more loosely connected collection of matter like a galaxy, it is possible for material at differing radii from the centre not to follow the same pattern as a solid disc. We would usually expect that after a time the galactic disc would settle down to rotating in a pattern where, moving away from the centre of the galaxy, after an initial fast growth, the speed tails off for stars that are further out. This results in a 'rotation curve' that is the standard behaviour predicted by Newton's law of gravitation.

Observing Andromeda, Rubin and Ford found that, unexpectedly, parts of the galaxy near the edge were rotating at close to the same speed as parts near the middle. The most obvious reason that this would happen was if, for some reason, there was a lot of matter distributed spherically around the outside of the galaxy, in what is known as a halo. (This can be a confusing term, as we usually think of halos as

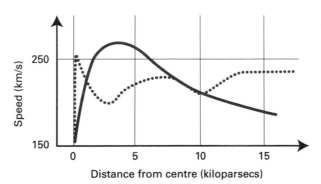

Velocity of stars in the Milky Way at varying distances from the centre. Solid curve is predicted by theory, dotted curve is observed.

Adapted from Creative Commons image CC BY-SA 3.0

being flat discs with a hole in the middle, but this 'halo' is more like a hollow ball.) Andromeda had been widely studied through telescopes – it has no such halo of visible matter. Like Zwicky, Rubin had found what appeared to be the signature of dark matter in its gravitational impact on the rotation of the galaxy.

Intrigued, Rubin and other astronomers began to collect data on other galaxies, finding a similar effect in each case. What's more, just like Zwicky's clusters, they found that the galaxies were rotating sufficiently quickly that they should have flown apart if they only contained the visible matter that appeared to make them up. Rubin estimated that there was in reality around six times as much matter in the galaxies than appeared to be the case from the visible material.

As the concept of dark matter gathered support, Zwicky's idea of using galaxies as lenses was roped in as an additional method to demonstrate that there was something unexpected present. As the gravitational lensing effect gets stronger, the results produced will vary. Just as the thickness of a lens will alter its focal length, a particularly high amount of mass in a galaxy will mean, for example, that it has more tendency to produce multiple images of the original source around the outside of the lensing galaxy. Using measurements of the impact of gravitational lensing provided further support for the existence of dark matter.

As yet, though, dark matter itself was a mystery. What was it made of? It couldn't simply be the traditional components of stuff – quarks and electrons – because they *do* interact electromagnetically. Even if it *is* matter that we can't see out in the depths of space because it is not glowing, such matter would not behave as dark matter appears to. So, what are the alternatives?

Dark what?

There have been many attempts to pin down just what dark matter truly is. There are two lines of thought that are primarily followed – that it is composed of familiar particles that are already part of the standard model of particle physics (but that don't behave like ordinary matter particles) or that it is made up of exotic particles that would require us to extend the standard model.

The main candidate from the standard model is the humble neutrino. This was a particle that was dreamed up in 1930, 26 years before it was ever directly detected. It was predicted that the neutrino would be produced when unstable radioactive elements undergo what is known as beta decay. The 'beta' bit refers to an electron being emitted. Originally, radiation had been divided into alpha, beta and gamma rays, where each type of radiation has very different properties. These emissions from radioactive atoms were later identified as helium nuclei, electrons and high-energy photons respectively.

Although we're familiar with atoms having electrons surrounding the nucleus, in beta decay it is the nucleus itself that gives off an electron. We now know that the electron wasn't there beforehand and a force of nature called the weak force converts a neutron in an atom's nucleus into a proton, giving off an electron in the process and transmuting the atom to a different element. The electron's negative charge counters the new positive charge in the nucleus. But in 1930, the Austrian physicist Wolfgang Pauli realised that the electron alone wasn't enough to balance out what had happened.

Although the electron dealt with the change in charge, other properties of the atom that had to be conserved would have changed. Specifically, the atom lost energy and underwent changes in momentum and spin.* Just as the electron was needed to balance out the electrical charge, another particle was necessary to carry away the lost energy and momentum, while balancing out the spin. The new particle would have no electrical charge, so Pauli called his hypothetical particle a neutron.

By 1932, though, English physicist James Chadwick had discovered the neutral matter in the nucleus was made up of uncharged particles with similar mass to protons, which he called neutrons. These clearly weren't the same thing as Pauli's hypothetical particles, which had practically no mass. The Italian physicist Edoardo Amaldi suggested to his compatriot Enrico Fermi that a good name for Pauli's particle would be the neutrino, the Italian diminutive for the name. Soon after, Fermi developed the full theory of beta decay and the name neutrino was cemented in place.

Neutrinos are elusive and odd. They are elusive because they hardly ever interact with matter. They aren't interested in electromagnetism (ring any bells?) so are invisible. The nuclear reactions in the Sun produce so many neutrinos that trillions pass through your hand every second – yet they have no effect because of that lack of electromagnetic interaction. They are so difficult to detect that, as we have seen, despite overwhelming evidence that they were there, it wasn't until 1956 that their detection was announced.

---

* Spin is a property of a quantum particle. The name is misleading as it doesn't imply rotation in the normal sense. A particle's spin can only be multiples of ½ in size, and when measured only comes out 'up' or 'down'. Before measurement it will usually be in a 'superposition' of the two states 'up' and 'down', with a particular probability for each state.

It might seem that detection of neutrinos would be impossible, but just occasionally they do interact with another particle. Neutrino detectors are usually based deep underground in old mines. This shields the detectors from other, more obvious particles, leaving behind the neutrinos, most of which pass through the Earth as if it weren't there. Inside the underground chambers, the detectors often consist of vast tanks of fluid – water or the dry-cleaning fluid tetrachloroethylene. Again, almost all neutrinos will pass straight through the detector, but a tiny number do interact with the particles in the fluid, either causing a change in a molecule that can be detected or producing tiny bursts of light which are picked up by sensitive detectors that surround the fluid.

By the 1960s it had been discovered that the neutrino wasn't a single particle, but came in three 'flavours', each associated with another particle: the electron and the other two electron-like particles, the muon and the tau particle. This helped explain odd results from early attempts to detect the neutrino flow from the Sun. The volume of neutrinos the Sun gives off can be estimated, but the early detectors suggested that only about one third of the expected number arrived at the Earth. These early detectors could only spot the electron neutrino and consequently didn't pick up around two-thirds of the possible arrivals.

Exactly which type of neutrino did arrive at any moment was itself variable. It turned out that neutrinos can undergo a process known as oscillation whereby they shift between flavours in mid-flight. Up to this point, most evidence pointed to the neutrino, like the photon, having no mass at all. But the only mechanism envisaged for neutrino oscillation required the particle to have a mass – though a very small one. To get an idea of scale, the electron – not exactly a heavyweight in particle terms – has around 4 million times the mass of a neutrino. (The mass varies slightly between flavours.)

Given that neutrinos do have mass, which means they have a gravitational effect, their lack of electromagnetic interaction made them an immediate candidate for the mystery particle of dark matter. However, there remained some problems. We have now got fairly good at detecting neutrinos – yet the detections don't suggest the kind of halo distribution required for dark matter to do its job. And there's also the matter of neutrinos being too hot.

The now-standard dark matter model refers to 'cold' dark matter. This term reflects the relationship between the speed of particles and the temperature of a body made up of those particles. The hotter, for example, a gas is, the faster the particles zip around. From the apparent configuration of the early universe, it seems that the dark matter particles should be slow-moving, otherwise they would not have been able to help structures like galaxies form – the particles would be moving too fast for gravity to bring them together. There's no such thing as a hot halo.

However, in reality, neutrinos have proved to be nippy little things, moving at very near to the speed of light. Briefly, even this seemed an underestimate. In 2011, scientists running an experiment called OPERA that sent neutrinos around 730 kilometres from CERN near Geneva to Gran Sasso in Italy announced that they had discovered that neutrinos travel faster than light. There was frenetic press speculation that this put Einstein's special theory of relativity, with its light speed limit, in danger. (For some reason, the press loves the possibility of proving Einstein wrong.) It turned out, though, to be an error caused by a faulty cable connection combined with an over-fast clock. Nonetheless, neutrinos do travel at close to the speed of light.

Being 'hot' doesn't entirely rule out neutrinos. It would have been just possible that larger structures of vast galactic clusters – big enough to capture even fast-moving neutrinos on a large scale – formed first, then broke up into smaller galaxies. To distinguish whether clusters or individual galaxies came first, we need to get a view of how the whole universe looked when matter first formed – and remarkably we have just that in something called the cosmic microwave background radiation.

The radiation from everywhere

Telescopes are time machines. As light travels at a finite speed (around 300,000 kilometres or 186,000 miles a second), the further away an object is, the further back in time we see it. When we look at the Andromeda galaxy, which is 2.5 million light years away, we are seeing it as it was 2.5 million years ago.

In principle, with a good enough telescope, the furthest we can look back is around 13.5 billion years in time to the point where the universe became transparent when atoms formed. Before then, matter was electrically charged and soaked up any passing light. The light that was zipping around the universe when it became transparent is still out there. At the time it would have consisted of extremely high-energy photons of light – gamma rays. But something has happened over the billions of years since that light was emitted. The universe has expanded.

Remember that it is space itself that is growing bigger. As it does, it will have an effect on the light that is passing through it. If you think of light as a wave, and the space it travels through expands, then effectively the wave will be stretched out like a concertina. This produces a longer wavelength, shifting the light to be more red.* And the longer that the light has been in motion, the more the universe will have expanded since it set off, resulting in bigger and bigger red shifts.

What was extremely high-energy light, with a very short wavelength, has gradually moved through X-rays, ultra-violet, visible light and infra-red all the way down to microwaves. These are a kind of light where photons have much lower energy than the visible – microwaves have more in common with radio than visible light – but we are familiar with microwaves because they happen to have just the right energy to get water molecules excited. This makes them great for heating up an item with water in it like a piece of food, hence the use of microwaves in ovens.

Back in 1965, two researchers at the Bell Labs facility in Holmdel, New Jersey were trying to use an aerial designed to pick up signals from the Telstar communication satellite to do astronomy. It had been discovered by then that stars don't just give off visible light, but also produce a wide range of the electromagnetic spectrum, including radio waves. Robert Wilson and Arno Penzias were looking for radio output from the edges of the Milky Way, but instead they found

---

* Light is a quantum phenomenon that can be treated either as a wave (probably more familiar from school), or a collection of particles called photons, as mentioned earlier. If you prefer to think of light as a stream of photons, the stretching of the space reduces the energy of the photons, again producing a red shift.

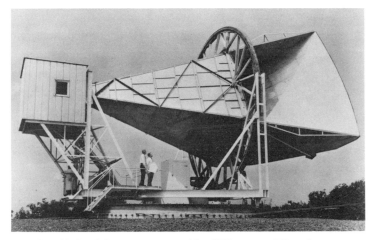

The Holmdel antenna used by Wilson and Penzias to detect the cosmic microwave background radiation.
NASA

a strangely uniform background hiss that came from each and every direction.

The signal was similar to the static picked up on old analogue TVs if they were tuned between stations – in fact part of that visual static and hiss was exactly the same signal that Wilson and Penzias received. For a while, the radio astronomers thought they were receiving earthbound interference. This is a common problem for radio astronomy – a faulty motor in a vacuum cleaner a few miles away can easily produce a false signal. But with careful analysis, Wilson and Penzias showed that the signal wasn't coming from any of their equipment, wasn't local and was just as strong whichever direction they pointed their antenna.

Another suspect for the cause of this mysterious signal was the droppings that were building up in their receiver because a family of pigeons were perching on the wide horn of the telescope (the droppings were euphemistically referred to when the experiment was written up as 'white dielectric material'). But even when they got rid of the pigeons and cleaned up the metal surfaces, that hiss remained. It was only when they talked to some other scientists who were looking for exactly such a signal that Wilson and Penzias realised just what they were picking up. The microwaves they were receiving were the remnants of the light that had been set free when the universe became transparent at around the age of 370,000 years.

This 'cosmic microwave background radiation' has been called the echo of the big bang. This is distinctly flowery and inaccurate language, considering the radiation is the remnant of something that took place a third of a million years *after* the big bang, but still it gives us an insight into the earliest view of the universe we can have. This radiation takes us back as far as light lets us see. In principle we could go further, though. At around one second after the big bang, the universe became transparent to a different type of particle we've already met: the neutrino.

A neutrino detector has already been used to make a very crude image of the Sun – the first neutrino telescope. As if to demonstrate just how little neutrinos care for ordinary matter, the picture was taken through the Earth, with the Sun on the far side from the detector used to create the image. But if we could ever make neutrino detectors work better, we might be able to detect this cosmic neutrino background radiation from the first second of the existence of the universe. Similar claims have been made for the even newer ability to detect gravitational waves in the structure of space itself. For the moment, though, we are limited to the view that microwaves give us of 370,000 years after the big bang.

When detectors on Earth are brought into play, like the crude radio telescope used by Wilson and Penzias, the cosmic microwave background is very smooth, producing identical levels from every direction. It is one of the reasons that the signal was identified as the remnant of the big bang's aftermath, because that radiation was expected to be the same from every direction. But in more recent times we have been able to study the cosmic background radiation in a lot more detail, uncovering minute variations in intensity.

This new view of the radiation pattern is thanks to three satellites: COBE, WMAP and Planck. COBE dates back to 1989, WMAP was launched in 2001 and Planck in 2009. Each satellite added in to the picture more detailed variations in the cosmic microwave background. The result looks dramatic in the images that were produced, but the level of contrast is hugely amplified. The actual variation is around 1 in 100,000, tiny changes from the constant background level.

**The cosmic microwave background
pattern from the Planck satellite.**
ESA/Planck Collaboration

When you look at the stretched-egg-shaped images from these satellites it is hard to make out just what you are seeing. It is thought that the pattern in the radiation is the result of the tiny variations in the makeup of the early universe that would result in galaxies forming. If this is correct, what we see is the equivalent of an ultrasound scan of the very early embryo of the universe – a truly remarkable picture.

This data enabled cosmologists to eliminate neutrinos as the source of dark matter. The view it gave of the early configuration of the universe did not suggest the formation, first, of vast structures that would later break up, which were necessary if dark matter were hot. Instead the universe seems to have accumulated structures from the small end of the scale – and for that to happen, cold dark matter would be needed.

More than this, though, the cosmic microwave background joined the behaviour of galaxies and galactic clusters as another piece of evidence for the existence of dark matter (or something producing a similar effect). In the very early universe, when the radiation first set off, there would still have been a lot of electrically charged (ionised) ordinary matter. This would interact strongly with the background radiation, but any dark matter present would not do so. The outcome would be different depending on the ratio of dark matter to ordinary matter.

This difference was predicted by running complex computer simulations of 'webs' of dark matter in the universe over time, part of a wider attempt to model the way the structures of the universe have been built up as it evolved. These were first run in the 1980s (when computing power was distinctly limited) by four astronomers, George Efstathiou, Simon White, Carlos Frenk and Marc Davis, sometimes known as 'the gang of four', and have continued to be improved up to the present day with the Illustris and IllustrisTNG projects.

The detected variations in the radiation bear a close resemblance to what would be expected if dark matter existed in the proportions now believed to exist from observations of galaxies and clusters.

MACHOs and WIMPs

With neutrinos pretty much ruled out as a candidate, the standout alternatives for a dark matter particle have been given the now rather dated humorous acronyms WIMPs and MACHOs. Respectively these acronyms stand for 'Weakly Interacting Massive Particles' and 'MAssive Compact Halo Objects'. (Justifying MACHO appears to have been quite a struggle.)

MACHOs provide what is arguably the most obvious of the solutions to dark matter – that it is just ordinary matter after all, but ordinary matter that you can't see, just as Fritz Zwicky first envisaged it. There's plenty of such matter around, from dust to black holes, after all. However, the evidence against MACHOs is strong.

One of the arguments for dark matter is that in the early universe there seemed to be more mass than there were ordinary elements to account for it – but if dark matter were just unseen ordinary matter, this imbalance would not occur. And it's hard to see why such dark matter would end up forming spherical halos rather than the discs of galaxies made from ordinary matter.

Despite this, attempts have been made to detect MACHOs outside the halo from the way they should act as gravitational lenses, bending light as it passes by, and there appear to be nowhere near enough dark conventional objects out there to account for the scale of effect from dark matter. There are also problems because black holes, which should account for a fair part of the MACHO contribution, can only form from a relatively large star – and the distribution of dark matter in many cases precludes such large concentrations of mass.

However, there could, in theory, be another type of black hole – a so-called primordial black hole. Instead of forming from collapsing stars, these would have come into being in the very early days of the universe when matter was first forming, when the violent fluctuations in spacetime after the big bang may have been capable of compressing matter to form black holes. These could be pretty much of any size, right down to micro black holes with a mass of a fraction of a gram.

By modelling different distributions of such black holes, it has been suggested that relatively light black holes, with masses typically between 0.06 and 1 times the mass of the Sun, could generate at least some of the effects attributed to dark matter. As a note of caution, though, a University of California, Berkeley study in 2018 analysed 740 supernovas to look for gravitational lensing from primordial black holes. Their finding: no more than 40 per cent of dark matter can consist of black holes, and probably none of it does.

As yet, then, there is no evidence that such black holes exist, or produce the effects of dark matter, but there are suggestions that at some point the LIGO gravitational wave detector could be used to detect mergers of such unusually small black holes. If it does, they could be back on the table in the ever-changing game of spot-the-dark-matter. For the moment, though, the MACHO particles have to give way to the WIMPs.

WIMPing out

At first sight, Weakly Interacting Massive Particles, which have been the prime candidate during most of the search for dark matter particles, provide little more than a label for the observed phenomena. After all, by definition dark matter is weakly interacting with ordinary matter, and to have a gravitational effect, the particles need to have mass. However, the theoreticians had a more specific idea in mind for that 'massive' aspect. A WIMP should not have a barely detectable mass like a neutrino, but should be more chunky, specifically around the mass of the Higgs boson.

The Higgs is the particle that caused a huge media stir for the Large Hadron Collider at CERN in 2012 and 2013. It has often been described as the particle that gives other particles their mass, but this is misleading. The theory that led to the search for the Higgs requires there to be an extra field in the universe, on top of something more familiar like the electromagnetic field. It is interaction with this 'Higgs field' that provides some of the particle masses we can detect. If there is a Higgs field, we would expect there to be Higgs bosons, which are disturbances in that field. And that is what was detected at CERN.

One of the reasons that it took a long time to pin down the Higgs is that the theoreticians couldn't tell the experimentalists where to look – they didn't know what mass the Higgs would have. However, the possibilities were eventually whittled down by elimination and it was discovered that the Higgs had a mass of around 125 GeV.* This is around 100 times more massive than the proton and neutron particles in the atomic nucleus, similar to a tin atom.

The idea that dark matter particles should be of around this mass is suggested by the processes that are thought to have occurred as the young universe cooled down after matter first formed. When there was lots of energy around, very heavy particles and their equivalent antiparticles (more on these below) could form, then return to energy as they annihilated. But as the universe cooled, there was less energy available at any location to produce high-mass particles.

From calculations based on the probable quantities of particles left behind with different masses, it would seem that the apparent density in the universe of dark matter particles required to produce the dark matter effect reflects a collection of particles with a similar mass to the Higgs boson.

Self-evidently from the Higgs discovery, the amount of energy required to produce particles of this mass is available from the Large Hadron Collider. However, despite years of searching, nothing has turned up in the collider that bears any resemblance to a dark matter particle, nor has anything similar turned up when studying cosmic rays (high-energy particles from outer space). In these experiments, the particle that is hunted is not detectable directly. Instead, the experiments detect other particles, generated from the energy of some interaction of the target particle.

These second-generation particles are produced by a mechanism that requires the equation $E=mc^2$ to understand.

---

* Giga electron volts, i.e. billions of electron volts. Particle physicists refer to the mass of particles in units of electron volts/c^2 (where c is the speed of light), but often just call them electron volts (eV), reflecting the energy in their mass.

As we have seen, Einstein's equation shows a direct link between energy (E) and matter with mass m. (The 'c' is the speed of light.) Energy can be converted into matter and matter into energy. The most common way for matter to become energy is for matter and antimatter to collide.

Despite its popularity with science fiction writers, antimatter, the power source of the USS *Enterprise*, does exist. Every matter particle has an antimatter equivalent with opposite properties, notably its electrical charge. When a matter particle and its antimatter twin come together, the matter is converted to energy in the form of photons of light. And, it's argued, the same should be the case for dark matter and anti-dark matter.

Admittedly there's a chain of 'if's here. If anti-dark matter exists, and if the energy generated by dark matter and anti-dark matter coming together then spawns a spray of ordinary matter/antimatter particles, we should be able to detect those particles. If we can then estimate what we should expect to see coming out of matter generated from the energy of ordinary matter/antimatter annihilation and there's actually more, perhaps it comes from dark matter.

Note that there is no suggestion that the dark particles would *be* Higgs bosons, or for that matter would be conventional atoms of this mass – each of these would interact with other matter differently from the elusive dark matter particles. Instead, the hope has mostly been tied in with a particle theory called supersymmetry, which predicts that all the known particles should have more massive partner particles.

In the very successful standard model of particle physics, particles divide into two types: fermions and bosons. Each broad type behaves very differently. Fermions, which can be thought of as matter particles, include quarks, electrons and their heavier equivalents, and neutrinos. These particles don't like to be too squashed up together: a rule known as the Pauli exclusion principle means you can't put two of the same particle in exactly the same state in the same place. The other set of particles in the model are called bosons. These are the particles that carry forces, including the electromagnetic force carrier, the photon, which doubles as the particle of light. Bosons are far more sociable than fermions – you can crowd as many as you like into the same space.

This might seem to include more than enough particles to go round, yet many theorists hope for far more. Supporters of string theory, which attempts to combine the otherwise incompatible quantum theory and the general theory of relativity, expect there to be the supersymmetric particles mentioned above. Their existence is essential for most variants of string theory to work. Supersymmetry says that every particle we know has an equivalent 'supersymmetric partner' of the opposite kind. Each fermion should have a boson partner (identified by putting an 's' in front of its name) and each boson should have a fermion partner with an 'ino' ending.

So, for example, a quark's supersymmetric boson partner is a squark and a photon's supersymmetric fermion partner is a photino. No supersymmetric partner particle has ever been found. Theoreticians love the 'beauty' of the structure, but there is no experimental evidence that it exists. However, if it *does*, there has been a suggestion that neutralinos, which are predicted to be the lightest of the supersymmetric partners, could be the particles behind dark matter.

Given the naming convention, you might expect a neutralino to be the fermion supersymmetric partner of a neutrally charged boson – and it is, but rather than being a partner of a single boson, the neutralino (strictly coming in four different varieties) arises from a mixing of quantum states of the supersymmetric partners of the neutral bosons: the photon, the Z particle and the Higgs, which are respectively the photino, the zino and the higgsino. Yet again (as is the case throughout string theory), there is no experimental evidence that supports this concept. Even so, neutralinos were attractive to theoreticians, particularly as calculations of how many neutralinos should have emerged from the origins of the universe match up well with the estimated mass of dark matter in the universe.

Despite attempts to hunt for the remnants of dark matter/anti-dark matter annihilation both from a dedicated satellite and an experiment on the International Space Station, nothing has been sighted. These experiments have told us something new – but only about conventional matter events in the universe. For example, the gamma ray observatory satellite Fermi detected unexpectedly high levels of radiation from the centre of our galaxy. But to use this kind of data to infer anything about dark matter is a like using an unidentified flying object to speculate about alien technology. The chances are that the UFO is actually a perfectly ordinary flying object that hasn't been identified yet – and to do much work on the assumption it can teach us anything about aliens is science fiction, not science fact.

It might seem that the failure of WIMPs to appear experimentally should be no surprise – after all, dark matter only interacts through gravity, and gravity is so weak that individual particles are never going to be detected this way. However, if WIMPs did exist with this kind of heritage, they should be produced in colliders and be detectable as described above from their decay products. They haven't been. Similarly, given that space should be full of these fleeting particles, then – like neutrinos – we would expect direct detections from experiments looking for their very occasional interactions with ordinary matter. Yet no so such detection has occurred.

The WIMP hunters

To be precise, evidence for WIMPs has never been effectively corroborated, but claims of detection have been made. A number of experiments have been constructed to spot these dark matter particles in the wild. Some of the detectors look for direct evidence of dark matter particles passing through the Earth. Like neutrino experiments, these devices are buried deep underground to prevent other particles being picked up by accident. Scientists then look out for unexpected collisions with otherwise non-interacting particles. The outcomes of these collisions are noted using either superconducting devices called SQUIDs or scintillation detectors.

SQUIDs (Superconducting Quantum Interference Devices) make use of a quantum structure called a Josephson junction which is usually used to detect tiny changes in magnetic fields. These devices are very temperature-sensitive – they have to be kept at an extremely low temperature to function. If a dark matter particle collision occurs nearby, it should generate a tiny amount of heat, which is enough to briefly disrupt the superconductivity.

Detectors that rely on scintillation pick up the generation of a tiny flash of light as a result of the collision giving energy to an electron in the fluid filling the detector. The electron then drops back to its usual energy, giving off a photon of light. The vast majority of these experiments (using either technology) have consistently failed to find anything, despite a few false alarms.

An early example was CDMS (Cryogenic Dark Matter Search) and its successor CDMS II, a spin-off from the Center for Particle Astrophysics at the University of California, Berkeley, an organisation that would later be involved in supernova research (see page 105). This experiment attempted to detect the tiny, infrequent interactions expected between neutralinos and atomic nuclei. The original detector, located 20 metres below ground at Stanford University, was simply not shielded well enough and picked up far too many ordinary interactions to provide any useful data. CDMS II (and its successor SuperCDMS), along with many other later dark matter detectors, ended up underground in an old mine (in this case, 750 metres down in Minnesota) to keep out as many unwanted natural particles as possible.

In 2007, the enhanced CDMS detected something: a pair of interactions over two months apart. Unfortunately, it was not possible to establish that these were WIMP interactions. One competitor, called DAMA, has produced a larger number of results. Since 1995, DAMA, a scintillation detector based 1,400 metres under the Gran Sasso mountains in Italy, has several times detected what observers have suggested could be dark matter particles – yet no one else can duplicate their findings. There was a potential loophole as DAMA was using a subtly different detector to its competitors – but a recent experiment, COSINE, run by a collaboration between the US, the UK and South Korea, located in South Korea and running from 2016, used an identical approach. As yet there has been no confirmation of DAMA's detections, making it unlikely that they establish the existence of dark matter.

If nothing else, this illustrates just how hard it is to eliminate false readings from these ultra-sensitive devices. Bear in mind that the equipment used in these experiments can't see the searched-for particles. All researchers have to go on is that some of the particles in the detector will be given a jolt. If everything else has been eliminated as a cause for this, it is likely to be because a dark matter particle has hit an atomic nucleus and produced a small burst of energy. But that 'everything else is eliminated' is not easy with spurious causes such as natural radioactivity in the surrounding rock, or our old friends, neutrinos.

As well as placing the detectors underground, ingenuity has to be put into both keeping the wrong kind of activity out and distinguishing dark matter in action. Sometimes the effort to provide shielding can seem quite bizarre. In one detector, lead from the hull of an old French galleon was used. Lead is good at stopping conventional radiation, and because this was old lead, it had had time to lose much of its own natural radiation, reducing the background noise it would generate. It was hoped that dark matter would be distinguishable from other radiation as it isn't distributed uniformly, so there should be periodic highs and lows as the Earth and the solar system move around their respective orbits (the solar system orbits the Milky Way galaxy) – but as yet these have not come to light.

In early 2019, a group at the University of California, Davis claimed to have discovered a possible explanation for DAMA's stubborn ability to make what appear to be occasional detections where no one else can do so with similar or more sensitive equipment. If true, like the spurious faster-than-light neutrinos (see page 40), the culprit would be a technical fault. The DAMA detector consists of

25 cylindrical scintillators with a device called a photomultiplier tube – effectively a light amplifier – at each end. The Davis team discovered that a small contamination of helium in the photomultiplier could cause an effect similar to the detections.

The DAMA readings do vary between summer and winter, which is suggestive of an external source like dark matter, rather than internal radiation. However, helium is produced by the radioactive decay of radon and other geological processes, which are also expected to have some annual variation, rather than a constant output. Several techniques have been suggested which would distinguish between a contamination and dark matter, but they have yet to be deployed.

At the time of writing (early 2019), among the latest detection experiments to announce results were XENON1T, like DAMA in Italy's Gran Sasso National Laboratory, and PandaX at the China Jinping Underground Laboratory in Sichuan province, both using scintillation in a chamber containing liquid and gaseous xenon. There has been unanimity: nothing found. XENON1T was able to announce a record low background level – effectively it had better eliminated alternative sources of events than its predecessors – making that absence of dark matter detection all the more stark.

One alternative approach that has been suggested for some time, but is now being treated more seriously, is to look for 'fossil' remains of dark matter impacts. The idea is that rather than sitting underground waiting for dark matter to strike, we should look at minerals which have already been underground for a long time and see if it's possible to detect a change in the material that was caused by the impact of a dark matter particle, which in some cases should leave a tiny track as a result of atomic recoil from the impact.

The idea of looking for fossilised impact residues underground dates back to the 1980s when experiments were carried out to try to pin down the impact of magnetic monopoles* – these failed. In the mid-1990s a similar approach was suggested to look for remnants of potassium atoms recoiling from dark matter impact in the mineral mica, which it was believed could be distinguished from similar impact from common or garden radiation. This was never tried, but was revived in a 2018 paper from researchers working in Sweden, the USA and Poland.

The new venture looked at other minerals: halite (a version of sodium chloride), epsomite, olivine and nickel-bischofite. As yet all the researchers have done is estimate the sensitivity of detectors using these materials, but whether they will ever be deployed is unknown – it may be that, as null results continue to come in from other detectors, the investment is considered unnecessary, even though in principle it may be possible to get greater sensitivity from this fossil remnant approach.

Although WIMPs remain the most popular candidates for dark matter particles, the lack of success in finding them, and the failure to date of MACHOs to match observation, has resulted in theoreticians dreaming up a more exotic alternative type of particle.

---

* Magnetic monopoles are hypothetical particles that have only one magnetic pole (north or south) – whereas all known magnets have both. Some theories, such as string theory, predict that monopoles should exist, but they have never been observed. The standard Maxwell's equations for electromagnetism assume that magnetic monopoles do *not* exist, but the equations can be extended to take in their existence should they ever be found.

Axion washes whiter

A leading alternative dark matter candidate is the axion. It sounds like a washing product for a good reason – it was actually named after a Colgate-Palmolive dishwasher detergent – but the axion particle is a favourite of a few theoreticians. These hypothetical particles had already been dreamed up to explain an oddity in quantum physics (it should be no surprise by now that there is no experimental evidence for them existing). If they did exist, they would have limited interactions, be of low mass (even lower than neutrinos) and there would be lots of them. But there are other aspects of dark matter behaviour where axions are distinctly problematic as a solution. Like neutrinos, they are likely to be too hot to fulfil the role, and some models suggest they would form structures that have not been observed.

It was hoped that the axion problem would be solved one way or another by the Axion Dark Matter Experiment (ADMX), established in the 1990s at the Lawrence Livermore National Laboratory in California, run by Karl van Bibber and Les Rosenberg. Their detection device was an extremely sensitive radio receiver. Axions, should they exist, would be too light to have much chance of being detected during a conventional interaction with matter. However, when traveling through a powerful magnetic field, they should produce a photon – and that could be detected.

The idea was to produce an intense magnetic field in a cavity that would act as a trap for radio-frequency photons. If these photons were generated by axions passing through the trap, they would build up over time to provide an extremely weak radio signal. By 1997, ADMX was up and running – but not getting anywhere. After that initial prototype the first phase of the experiment ran through to 2004 with no detections. The second phase, which was supposed to be definitive, but detected nothing, was generally considered the final chance for the axion as a dark matter candidate, though this hypothetical particle does still turn up in some theoreticians' ideas about nature in general.

Whether we're talking WIMPs or axions, attempts to come up with a distinctive dark matter particle do, however, seem to make a rather strange assumption.

A proliferation of dark

We have quite a good model of how the ordinary matter and forces in the universe act. As we have seen, the central description of every 'normal' particle, called the standard model, is composed of a family of seventeen particle types (though supporters of 'supersymmetry' believe there should be at least another seventeen) making up most of observable reality. (See Appendix for a diagram of the standard model.)

It seems, then, quite a jump to make the assumption that dark matter, comprising five times as much of the stuff of the universe as ordinary matter, is made up of a single type of particle.* We have no reason to suspect this, other than to keep it simple for theoreticians (and the universe is not particularly known for this trait). It's entirely possible that dark matter consists of a similarly large family of particles, rather than a single culprit.

---

* As a reminder, 27 per cent of the universe is dark matter and 68 per cent dark energy, leaving only around 5 per cent as everything that we directly observe.

If we see the picture in reverse, imagine existing in a dark matter universe and trying to predict what 'ordinary matter' was made of. Taking this approach, we might assume there was a single 'ordinaryon' particle that makes up everyday stuff – far from the truth. The physicist Lisa Randall has called those who take the single particle approach 'ordinary matter chauvinists', because of the assumption that somehow our matter should be much more complicated than primitive dark matter.

We could even envisage an unseen parallel dark matter universe, in which 'dark light' shines from dark suns onto dark planets occupied by dark beings. In reality, this picture is unlikely, as the way dark matter behaves suggests it has little interaction with itself other than gravitationally – limiting the interest value of dark matter worlds – but it's still a fun piece of speculation.

Though the full 'dark universe' picture is extremely unlikely, some physicists have pondered how a 'dark photon', a sort of photon with mass, would behave. This would be an extra particle in the standard model that was a 'gauge boson' – it would be a force-carrying particle that provided interaction between dark matter particles which themselves had a very light mass. In the unlikely event that such a particle existed, in principle it might cause tiny displacements of the mirrors used in gravitational wave detectors such as LIGO.

Gravitational waves* are themselves discovered as a result of incredibly small movements of these mirrors – shifts

---

* Gravitational waves were predicted by Einstein in 1916 and first observed in 2015. They are travelling contractions and expansions of the fabric of spacetime itself, caused by major gravitational events, such as colliding black holes. For more information see the author's *Gravitational Waves* in this series (Icon Books, 2018).

smaller than the size of a proton. If a stream of these 'dark photons' were to pass through the detectors, it is possible that the movements caused could be picked up and distinguished from other vibrations. It is more likely, should they exist, that such detection could be made with the proposed, much larger space-based replacement for LIGO called LISA. But it ought to be stressed that this possibility piles hypotheticals on hypotheticals and could well be seen more as a justification for extending the sensitivity of gravitational wave detectors than anything with a significant likelihood of success.

Another potential candidate for 'dark radiation' is the so-called sterile neutrino (itself highly speculative at the time of writing). Although neutrinos have been pretty much ruled out as the 'stuff' side of dark matter, an experiment at Fermilab in America called MiniBooNE, which was set up to study the way neutrinos change flavour, gave hints in 2018 that an extra flavour of sterile neutrinos could exist, as muon neutrinos appear to switch through an unidentified state before becoming electron neutrinos.

Just introducing the sterile neutrino would need a significant extension of the standard model of particle physics, as the 'sterile' part indicates it is not affected by the weak force, unlike ordinary neutrinos. While in principle, should it exist, the sterile neutrino could be a dark matter force-carrier, we are a long way from anything certain here – but it's another avenue of exploration.

Under our feet

A question that quite frequently gets asked of physicists working in the field is why dark matter isn't obvious

– specifically, why it isn't accumulating under our feet. The Earth, after all, was formed by gravitational attraction from clouds of dust and gas. So why hasn't it also accumulated more dark matter than normal matter, rendering the Earth far more dense than it actually is?

It's certainly true that all the detection experiments assume that there is a lot of dark matter passing through the Earth all the time, but there are two good reasons why the Earth wouldn't be overloaded with dark matter.

One is that, though there's theoretically a lot of dark matter about, it wouldn't have accumulated in the disc of the solar system (and then into the Earth) like normal matter. With only gravitational attraction, dark matter is expected to come together to form the terribly misnamed halos – hollow spherical-shaped bodies. This would mean that there was less available when the Earth formed than ordinary matter, despite there being more overall in the universe.

Secondly, if we imagine a particle of ordinary matter heading towards the Earth (ignoring its interaction with the atmosphere), it would accelerate due to gravity until it hit the surface. Here, electromagnetic forces would bring it to a stop, allowing it to become part of the Earth. A dark matter particle would have a similar gravitational acceleration, but lacking that electromagnetic interaction it would just carry on accelerating until it had passed the centre of the Earth, when it would be decelerated and would leave the other side with whatever initial velocity it had. It would be very unlikely to be moving at just the right speed and direction that it could be captured. It has been estimated that the amount of dark matter in the Earth is only a few grams.

Tout le MOND

In 1983, astrophysicist Mordehai Milgrom put the cat among the dark matter pigeons by coming up with a mechanism that would explain some of the effects ascribed to the never-observed substance without the need to introduce a new type of matter and potentially rewrite the standard model of particle physics. Milgrom, born in Romania in 1946, has spent his working life in Israel. His idea has become known as Modified Newtonian Dynamics or MOND for short.

The idea is simple but powerful. The effects that are ascribed to dark matter would mostly also be observed if gravity itself acted ever so slightly differently on the scale of galaxies and clusters than it does on more familiar scales of planets and stars. It's only an assumption that gravitation behaves exactly the same on all scales. The tweak required to Newtonian predictions would be small, but would enable fast-rotating galaxies to keep together when they apparently should fly apart, without the need for any extra stuff to be present.

We are used to thinking of gravitational effects as universal. However, we know that other physical behaviours are scale-dependent. After all, matter acts very differently at the level of tiny particles such as electrons and atoms than it does with familiar objects such as people and tennis balls. And it has already proved necessary to make small adjustments to Newtonian dynamics to deal with the general theory of relativity – which has mostly produced the same results as Newton's theory, but with subtly different outcomes in particular circumstances.

Milgrom's idea was initially seen as a significant

challenger to the existence of dark matter, were it not for the Bullet Cluster.

The silver Bullet

Among aficionados of dark matter, perhaps the best evidence that it exists comes from a galactic structure called the Bullet Cluster (so called as, with a lot of imagination, it looks a little like a stop-motion picture of a bullet trailing the gases that propelled it from the gun barrel). The Bullet Cluster appears to be the aftermath of two or more galactic clusters merging. The shape of the cluster has been described as a blob of a central region with a pair of bulbous outer regions like Mickey Mouse ears (though the regions are three-dimensional, not flat).

It's hard to see how this unusual structure would have resulted from the collision of clusters of ordinary matter alone. But the suggestion is that as the original clusters merged, the ordinary matter collided in the centre, giving off energy in the form of high-energy light. This loss of energy meant that the ordinary matter stayed around the central point of the collision. But dark matter doesn't interact through electromagnetism, even with itself, so it passed through that central region and kept going. Over time, gravity slowed it down, but not before it had ballooned out on either side of the central region.

The result was to produce two outer lobes primarily formed of dark matter, but capable of attracting enough ordinary matter to be visible, along with the central segment primarily formed from ordinary matter. This is certainly a possibility, and is the reason that the Bullet Cluster is beloved of many cosmologists and astrophysicists. It is considered (in a happy coincidence) to be the smoking gun in support of dark matter – and to have signalled the end of MOND.

Many physicists and science writers assume that the Bullet Cluster has finished the argument, much as the battle between the big bang and steady state theories of cosmology was finished off by the discovery of the cosmic microwave background radiation. However, things aren't that simple. Supporters of modified gravity have fought back and are winning higher levels of support than ever before.

MOND strikes back

The reality is that, while it's true that the basic version of MOND struggles to explain the Bullet Cluster, there are also rather more cosmological phenomena for which dark matter simply doesn't work well as an explanation. In these cases, it makes far more sense if we dispense with dark matter and move to a modified gravity theory. A classic example is the galaxy NGC 1560. Discovered way back in 1883, this spiral galaxy is in the region of 10 million light years away from us.

The rotation curve of this galaxy is way flatter than that expected from Newtonian theory – so it certainly needs something to explain it. And MOND's predictions fit the curve with beautiful accuracy. But the predicted curve from dark matter simply can't be made to reflect reality. All the evidence is, then, that the Bullet Cluster is no silver bullet for dark matter. And it certainly is no more convincing than examples like NGC 1560, which provide equally good data where only a modified gravity approach seems to work.

Other, more recent modified gravity approaches, such as Scalar-Tensor-Vector-Gravity (STVG), which adds an extra field interacting with ordinary matter, seem to have no problem with the Bullet Cluster. In fact, with a different cluster, the impressively named Train Wreck Cluster, dark matter seems not to work as an explanation, where STVG does.

Another apparent MOND-killer from those determined to get modified gravity set aside is a dwarf galaxy with the uninspiring name NGC 1052-DF2, which was discovered in 2018. What raised astrophysical eyebrows here is that this dwarf galaxy was rotating in such a way that it appeared not to have any dark matter in it. This is unlikely, but possible if dark matter exists. The general assumption in the theory of dark matter is that all galaxies required it initially to be able to form in the timescale they have had since the origin of the universe. But given that most of the dark matter is on the outside of a galaxy, the dark matter could be stripped away by interaction with other galaxies. As this dwarf galaxy is a satellite of a massive galaxy, such a process is theoretically capable of occurring.

However, if dark matter *didn't* exist and the usual behaviour of galaxies was down to some kind of modified theory of gravity, then it should not be possible to have a normal galaxy without this effect. This discovery of NGC 1052-DF2's atypical behaviour was initially trumpeted as another death knell for modified gravity – but once again, things aren't that simple. While this kind of implication could be true of the basic concept behind modified gravity, a full implementation of it would inevitably have extra complexities that could easily deal with this kind of oddity – as witnessed by the way that more detailed theories have dealt with the Bullet Cluster.

As it happens, with more analysis, although it would be at the extreme of what basic MOND predicts, the behaviour of NGC 1052-DF2 is not actually different enough from the norm to present a challenge to even basic modified gravity. This is because there is only very limited data available, and depending on how that data is analysed, the galaxy can be made to fit with MOND. There is no suggestion that those involved in using this galaxy to sound the death knell for modified gravity were cherry picking, but they certainly chose the interpretation of the data that was most likely to support their argument.

At the time of writing, this discovery was at best borderline, with a second group asserting that the data had been misinterpreted and NGC 1052-DF2 did not lack the effects of dark matter at all. Their proposal is that the distance to the galaxy had been incorrectly measured (an idea boosted by the recent realisation that the galaxy had been 'discovered' twice before, providing more data for the calculation). If, as this information makes possible, the galaxy is far closer to us than was originally thought, it would mean that the mass calculation was incorrect and the assumption that the galaxy was devoid of dark matter was false. The jury is out.

Emergent gravity

MOND (and its more recent variants) is not the only game in town when it comes to modifying gravitational theory to account for the apparent action of dark matter. A recent entrant to the game is known as emergent gravity (also called entropic gravity).

Our current touchstone for understanding gravity is Einstein's general theory of relativity (more on this to come). The theory has stood the test of time very well, except for its inability to be combined with quantum physics. Any attempt to bring them together results in impossible infinities cropping up. Some attempts to combine the two, notably loop quantum gravity, have to modify general relativity to enable spacetime to be quantised.* The newest such approach, emergent gravity builds on an observation that in some ways general relativity resembles thermodynamics, the physics behind heat and the motion of gas molecules.

In this model, rather than gravity being a truly fundamental force, it is 'emergent' as a result of entanglement (a well-studied quantum effect) between the quantum particles of spacetime. Emergent properties are common in nature, where a combination of apparently simple constituents produces a more complex whole. Everything from the structure of sand dunes and the form of snowflakes to life itself is described as emergent.† But it would be quite a step to make gravity emergent.

At the end of 2016, Dutch theoretical physicist Erik Verlinde of the University of Amsterdam showed that if emergent gravity existed in a universe with a positive cosmological constant (a universe like ours – see page 115 for more on the cosmological constant), then it would result in some deviation from the general theory of relativity

---

* This implies that, rather than being continuous, space and time are broken up into extremely small chunks.
† Think about it: as a human being, you are made of trillions of living cells. No individual cell can do much on its own: human abilities are emergent properties from the combination of all those cells.

– specifically that it would produce a push on matter that should have a similar effect to that ascribed to dark matter. As a neat bonus, the effect is predicted not to occur in high-density systems such as solar systems, only in more diffuse systems (with lots of empty space) such as galaxies.

However, there are significant problems. All the implications and detail of this approach have not been worked out, relying as it does on extrapolation from an idealised situation that is far simpler than reality. Even so, it has already been shown to be less effective than MOND in predicting the way matter rotates around galaxies, and worse than dark matter at explaining the behaviour of clusters. And it can't as yet provide any insights into very large-scale structures or the formation of structures in the early universe.

Nonetheless, given that this is a very recent and not fully developed theory, it is generating some interesting ideas and adds extra weight to the need to give modified gravity theories more consideration.

Gooey gravity

One final example of an alternative solution to the dark matter problem is one that makes dark matter a real substance, but matter that doesn't behave like a conventional collection of particles that are interacting through gravity. This is superfluid dark matter.

The best-known superfluid is liquid helium, where the element is cooled to below 2.17 K (that's 2.17 kelvin, units the same size as degrees Celsius, above the lower limit of temperature, absolute zero at –273.15°C). The atoms in the liquid helium become linked at the quantum level, turning it into a substance with no viscosity and with the ability to transmit thermal energy perfectly with no losses. That 'no viscosity' part means that if you start a ring of superfluid rotating it will continue to move for as long as it remains at a low enough temperature.

If dark matter were a superfluid, specifically a widely studied form of superfluid known as a Bose-Einstein condensate, rather than a conventional collection of particles, it would explain the failure to detect dark matter particles – as dark matter would not exist with its special properties in individual particle form. In effect, superfluid dark matter is the result of a type of combination of traditional dark matter models and modified gravity.

The attractive aspect of this approach is that it's possible to conceive of dark matter being in superfluid form on the scale of galaxies, but behaving differently on other scales. The physics suggests that in galactic clusters dark matter would dominate, but in non-superfluid form, while on the scale of a solar system, normal matter would overwhelm the effects of dark matter, making a better match to most observations than either simple dark matter or modified gravity manage on their own.

Like simpler modified gravity approaches, superfluid dark matter is as yet extremely hypothetical, but it's a new approach that is gaining some interest and could still represent the kind of breakthrough astrophysicists are looking for. In effect, the superfluid model provides modified gravity when the material is acting as a superfluid and conventional dark matter actions when not – giving it a 'best of both' approach. It's too early to tell if it will stand the test of time.

Galactic scaffolding

Whatever dark matter is, in reality – whether it's a real collection of stuff that only interacts through gravity or a tweak to the equations of gravitation for large-scale structures – it is one of the contributors to us being here at all. In the early days of the universe there was a lot of pressure from radiation, making it hard for the relatively weak force of gravity to pull together matter on a large enough scale to form galaxies. If there had been no dark matter or modified gravity, there simply wouldn't have been time for the large-scale structures we see in the universe – including our own Milky Way – to form. Because the gravitational impact of dark matter is the equivalent of five times that of ordinary mass, and it isn't affected by radiation, the result of the dark matter effect was a sufficient pull of gravity to enable these structures to coalesce.

Dinosaur disaster

One of the more entertaining theories relating to dark matter is that it could have been indirectly responsible for the destruction of the dinosaurs. Although it's not widely supported, this is not a crank concept, devised as it is by physicist Professor Lisa Randall and her colleagues.

For a long time, the demise of the dinosaurs was a puzzle. How could animals that had dominated the Earth for millions of years quite suddenly be almost all wiped out? (Not entirely so, in practice, as birds are the living descendants of some dinosaur species.) The explanation was the result of a wonderful bit of scientific detective work by the father and son combination of physicist Luis Alvarez and geologist Walter Alvarez.

In the 1970s, Walter was studying a particular layer in the Earth's crust known as the K-Pg boundary. This was the thin layer of material laid down in the period we identify as the transition from the Cretaceous period to the Paleogene (previously known as the boundary between the Cretaceous and the Tertiary before the geological period naming was revised). This layer dates back roughly 66 million years to the time when the dinosaurs disappeared in a mass extinction that also involved many other groups of organisms.

Working with his father Luis, Walter became interested in the level of the element iridium present in the K-Pg layer. This heavy metal is scarce near the Earth's surface as it is dense enough to have largely been pulled deeper into the planet. As a result, iridium in this layer (which was on the surface at the time it was laid down) is primarily extra-terrestrial, coming from the impacts of meteors. The original intention of Walter's work had nothing to do with dinosaurs. The idea was to get a feel for how long it took to lay down the layer by the amount of iridium that had been deposited, assuming it would arrive from space at a fairly steady rate. Instead, he and his father found a startlingly high quantity of iridium, around ninety times as much as would be expected from the 'background' arrival of the element.

A similarly elevated level of the rare metal was found in the same K-Pg layer at locations around the world – in some cases with as much as 160 times the expected quantity. It was estimated that to produce such an effect would require around half a million tonnes of iridium to arrive at the Earth all at once. The only obvious solution was an asteroid or comet impact, from a huge body, around ten to fifteen kilometres in diameter.

The theory rapidly took hold as the cause of the dinosaur extinction when more data came in, pushing aside alternatives such as the effect of volcanoes, which wouldn't have produced the same kind of deposit. When this vast object struck the Earth at high speed, apart from wiping out vast swathes of the surrounding territory with its shock wave, earthquakes and tsunamis, it would have created planet-circling clouds of ash and dust. First the hot debris from the collision would have caused worldwide fires and soaring heat levels, destroying habitats – then the screening from the Sun would have seen temperatures plummeting for years.

The final piece of the jigsaw took over a decade to find – the crater left behind by the impact. It might seem that an estimated 200-kilometre-wide crater would be easy enough to spot, but the impact wasn't necessarily on land and the crater would have been partially filled in during the 66-million-year interval since it was formed. It was eventually discovered that a likely culprit had already been found during oil exploration, but had not been widely covered in the press. It is the Chicxulub crater, half on land and half in the sea off the Yucatán peninsula in Mexico.

The destruction of the dinosaurs by an extra-terrestrial invader makes a great story, but where does dark matter come into it? The impactor itself was a perfectly normal asteroid or comet – not itself dark matter-related. However, Randall's theory is that dark matter could have been responsible for the impactor hitting the Earth. The suggestion is that as well as the spherical halo of dark matter surrounding it, the Milky Way has some dark matter in a disc that roughly parallels the disc of our galaxy. As the Sun passes around its orbit of the Milky Way, it doesn't stay exactly in the plane of the galactic disc but gradually moves up and down a little above and below the plane.

This means that, should such a plane of dark matter exist, the Sun would be dipping through it – and as it did so, the motion should produce gravitational effects that could disturb the orbit of an asteroid or comet in the solar system sufficiently to send it on a collision course with Earth. It's all very hypothetical, of course. Even if such a disc of dark matter existed, it wouldn't necessarily be the cause for a change in orbit. And the very existence of the disc is a topic of significant dispute.

As we have seen, in the conventional model of dark matter, it doesn't naturally form a disc like ordinary matter, but rather produces a spherical halo, as dark matter particles won't be interacting with each other electromagnetically. But Randall suggests there could be different types of dark matter particle (see page 61), and at least one of these types could have properties that enabled this kind of inter-dark matter interaction.

It might seem that this idea was devised to sell Randall's book on the subject – after all, who doesn't love cosmology and dinosaurs? However, it is a serious theory. Other astrophysicists are doubtful, due to the lack of confirmatory evidence, and because it requires an extension to the model of dark matter for which there is little current support. Whether you agree with it or not, it's an interesting reflection of the way dark matter's influence has spread beyond the simple requirement to explain the rotational behaviour of galaxies and galactic clusters.

No dark matter? No problem

Although the mainstream battle for the ability of apparently overweight galaxies to remain stable is still between dark matter and modified gravity theories, there is a third way – another option to explain this behaviour. In many ways it's the simplest possibility, and it comes not from a physicist or an astronomer but from a mathematician. Donald Saari was professor of mathematics at Northwestern University and until recently the director of the Institute for Mathematical Behavioural Sciences at the University of California, Irvine. He has suggested that dark matter could be a chimera.

Like the mythical beast, the suggestion is that dark matter could be simply in the imagination of those who went looking for it. This might seem unlikely, given that there has to be some cause for the unexpected rotational behaviour of galaxies. But Saari suggests that there is a very simple explanation – galaxies are behaving just the way that our current theory of gravitation predicts they should, but the astrophysicists have got their sums wrong.

Up until now we have casually taken for granted that it is possible to calculate the dynamic behaviour of a galaxy – a structure consisting of billions of stars. That's a big assumption, when you realise that the largest number of bodies acting together gravitationally for which we can make an exact calculation of their motion is two. So, for example, given just a single star and planet in the whole universe, we can calculate their gravitational interaction perfectly. However, add in one more body – a second planet or a moon orbiting the planet, for example – and the situation becomes too complex to fully calculate, as the gravitational pull and motion of each body interacts with the others. The result is chaotic.

This doesn't mean that it's impossible to do calculations and make predictions with more than two bodies. Astronomers have done well at making predictions of the motions of the planets and other bodies in the solar system for centuries – and it can now be done extremely accurately. But this is done using an approximation that gets closer and closer to the actual value, rather than an exact, perfect solution.

Now multiply that problem up for a whole galaxy. Take that most familiar of galaxies other than our own, the magnificent Andromeda galaxy. Andromeda is estimated to contain around a trillion stars and is around 220,000 light years across. We don't know where every star is, how it is moving or how all the other stars in the galaxy influence it. We can only deal with the galaxy's motion by drastic approximation. And it is possible that there are flaws in the mathematical method used to do this.

The approach taken is to consider the galaxy as a continuous piece of material (clearly this is a severe approximation). For this to work it is assumed that each of the stars in the imagined continuum has its gravitational acceleration towards the centre of mass of the whole. Just like any orbiting body, a star doesn't plummet into the centre because there is also rotation. Any particular star is considered to be in a (calculable) two-body relationship with the imagined continuum of all the stars that are closer to the galactic centre. The rest of the galaxy is ignored, as its effects should cancel out.

Saari points out that this approximation method has issues. Imagine zooming in on just two stars in the galaxy, orbiting the galactic centre near to each other. Bearing in mind gravity's dependence on the inverse square of the separation (the closer the bodies, the greater the attraction), their closeness means that the gravitational interaction between the stars will dominate compared to the interaction between one star and the rest of the imagined continuum. (Something similar happens in the solar system when bodies in orbit around the Sun get too close to Jupiter, for example.) The result will be that the assumption of acceleration towards the centre of mass will be violated and the faster of the two stars will pull the slower star along with it.

To quote Saari, because the equation typically used 'cannot handle even simple, well-behaved systems of discrete bodies, one cannot expect it to reliably predict what happens with more complicated settings'. And there's no doubt most galaxies fit that 'more complicated' description. Perhaps surprisingly, Saari does expect dark matter to be discovered, but not in the distributions currently predicted. He concludes: 'Indeed, it appears that much of the search for massive amounts of dark matter is a search for something that does not exist; it is a mathematical error. On the other hand, with so much going on in our heavens, one must expect that something will be found that can be labelled "dark matter".'

Quite where this final expectation comes from isn't clear – perhaps it is an attempt to prevent it seeming that astrophysicists have wasted their time on dark matter theories. However, Saari's mathematical challenge adds to the complexity of the picture of just what is happening in the universe. Not only could dark matter be down to unknown particles or a modification of gravity, it might be a mathematical illusion.

Such is the uncertain position of dark matter at the moment. But what about our other dark property, dark energy? To find out more about that, we first have to see how discoveries were made about the size and the expansion of the universe.

HOW BIG IS THE UNIVERSE?

4

Universal measurements

We have been speaking of galaxies and clusters as if they were obvious things, but most of us have never seen them outside of photographs. Pinning down just what was out there and being able to get a realistic idea of the size of the universe proved tricky even once telescopes had been added to the astronomers' armoury.

It's true that some stars are brighter than others, and a naive approach would be to say the brighter the star is, the closer it is to us. We know the Sun is a lot brighter than the other stars, so this sort of makes sense. But brightness could never be used as a sufficient measure of distance. The stars might all be the same distance away, but differing in brightness (as assumed in Aristotle's model) – or they could be scattered across a wide range of distances. And the same could apply to what we now know to be distant galaxies. But how do we measure distances in space? Without being able to go out with a tape measure, how is it possible to find out?

The earliest practical way used to measure distances to the stars can be seen in action with a simple experiment. Hold a finger up in front of your face and alternately close your left and right eyes. As you do so, the finger seems to move against the background. Now hold your finger at arm's length and repeat the experiment. It still appears to move, but not as far. The further away something is, the less it seems to move when you switch viewpoint between your eyes.

You should be able to use the same technique, called parallax, to tell how far away a star is. Look at it through one eye and then the other and it should appear to shift. But, in reality, if you look at the stars and switch eyes there is no visible movement. It's not surprising, given how far away we now know that they are. But imagine your eyes were 300 million kilometres apart. Then the shift would be much bigger. This is relatively easy to do. If you look at something in the night sky, then wait six months and look again, the Earth will be on the opposite side of its orbit around the Sun, shifting sideways by about 300 million kilometres. And this is indeed enough to measure the distance to the nearer stars.

Astronomers make use of this effect in a unit of measurement. The distance to even the nearest star is quite messy using the standard scientific measure of distance, metres. Proxima Centauri is around 38,000,000,000,000,000 metres away. Admittedly scientists deal with much bigger values than this, by using exponentials – powers of 10. A scientist would write this distance as 3.8×10^{16} metres – where 10^{16} means 1 with 16 zeroes after it. (If they wanted to show a very small number like 0.00000000009, the would write 9×10^{-11}, meaning 9 divided by 10^{11}.) Astronomers, however, tend to talk to common folk in light years and to themselves in parsecs.

As we have seen, a light year is the distance light travels in one year – around 9,467,000,000,000 kilometres. This is useful, because just a few light years can represent a good-sized distance in space, and the value instantly tells us how far back in time we are looking. When, for example, we look at the Andromeda galaxy, at around 2.5 million light years distant, we are seeing it as it was 2.5 million years ago, long before human beings even existed.

The parsec comes back to our trick with the moving finger – parsec is short for 'parallax arc second'. A full circle is divided into 360 degrees, each of them a degree of arc, of curvature. So, one degree is 1/360th of a complete circle. A minute of arc is 1/60th of a degree, and a second of arc is 1/60th of a minute. If an object is one parsec away, then the movement seen in it when the Earth goes from one side of its orbit to the other is one second of arc. By doing a little geometry, given the diameter of the Earth's orbit, you can work out that the distance from Earth of a star that moves by one second of arc is around 3.1×10^{16} metres, which makes it about 3.26 light years.

After the initial excitement of measuring the distance of our near neighbours using parallax, it soon became apparent that the universe was much bigger than anyone expected. There are plenty of stars that don't undergo any detectable shift at all, even from the opposite sides of the Earth's orbit – they are too far away. To measure their distances involves a little more guesswork, and the use of standard candles.

By standard candlelight

The idea is simple. If I have two identical candles and position one further away than the other, the more distant candle will look dimmer than the nearer one. If I measure the relative brightness of the two and know the distance to the nearer candle, I should be able to work out the distance to the more distant one. Similarly, if there are two stars in the sky that have the same actual brightness, but one is further away, we can measure the distance to the closer one by parallax, and use the difference in apparent brightness to work out how far away the dimmer one is.

That's all very well, but how do we know that two stars have the same actual brightness? Maybe the dimmer one is the same distance away, but just ... dimmer. Or a brighter star could be further away than a dimmer star it is being compared with, making it appear to be giving out less light than it actually is. In the constellation of Orion, for example, the brightest star Rigel is three times further away than the dimmest of the main stars, Mintaka, which is the right-hand one of the belt. (Mintaka is actually four stars in a complex system.)

To be able to use standard candles, astronomers had to find a way to identify particular types of star that could be relied on to have the same brightness. Luckily there are distinct star families where the brightness can be predicted with some accuracy. Different types of stars have varying combinations of material in them, which can be identified using a spectroscope. And one kind of star is even more distinctive than the rest.

The first standard candles were variable stars. These are stars that get brighter and dimmer over time, following a regular pattern. The classic variable stars used as standard candles, called Cepheid variables after the constellation Cepheus, seem to be in a cycle where sometimes they are puffing up with pressure from the reaction in the star, and sometimes they are collapsing under the force of gravity.

The cause for the variation was not known when measurements on their variability were first made by a pioneering female astronomer, Henrietta Swan Leavitt. A graduate of Harvard, Leavitt worked at the College Observatory, studying variable stars in a pair of satellite galaxies to our own, known as the Small and Large Magellanic Clouds. She catalogued a large number of variable stars and discovered that the brighter stars appeared to have a longer period of oscillation.

In a paper from 1912, Leavitt established that there was a straight-line relationship between the period of a group of these stars in the Small Magellanic Cloud and their observed brightness. As all these stars were at roughly the same distance, she deduced that the time they took to brighten and dim could be used as a measure of their actual brightness. So, if you find two of the same kind of variable stars with the same rate of flashing but one is always brighter than the other, the brighter one is closer. One year after, the distance to several of these Cepheid variables in the Milky Way was measured by Danish astronomer Ejnar Hertzsprung, calibrating these 'standard candles'.

By the 1920s, the positions of a great many stars had been measured – but one thing remained uncertain. Was the Milky Way the entire universe – was every star in the Milky Way or floating around its environs – or was this vast collection of stars just one of many 'island universes' – galaxies, spread across a much bigger universe? No star had yet been discovered outside the span of the Milky Way, which by then had been established to be around 100,000 light years from side to side, or in what amounted to extensions of it, like the Magellanic Clouds.

In 1923, an American astronomer called Edwin Hubble came up with the definitive evidence. He was studying the Andromeda galaxy, then known as the Andromeda nebula. Under the powerful telescopes available by then, the nebula was revealed to be a vast collection of stars. And Hubble spotted the right kind of variable star – a Cepheid variable – to be able to pin down how far away the nebula was. He calculated that it was 900,000 light years distant – far outside the Milky Way.

In fact, Hubble made a mistake in his distance measurement. There are two types of very similar Cepheid variable star, with significantly different brightness for the same rate of flashing. He was comparing a star of one type with another star from the second family. Once this mistake was corrected for, it was found that the Andromeda galaxy was nearly 2.5 million light years away. As we have seen, this makes it the most distant thing visible with the naked eye.

Hubble's discovery that the Milky Way with its billions of stars was just a small part of the universe would have been an impressive enough achievement – but he went on to discover something even more amazing.

It's getting bigger

Hubble's second great discovery was a result of measurements that had been made for a number of years using a spectroscope. As we've seen, this is the technology that astronomers use to work out what elements are inside a star. In fact, it's rather remarkable that it is much easier to tell what elements make up a star than it is to discover exactly how far away it is.

When you heat things up, they start to glow. But rather than give off every single colour of the rainbow, different chemical elements produce light in very specific bands of colour. The metal sodium, for example, has a strong yellow band, which is why sodium vapour street lamps give off their distinctive yellow-orange glow. But this process works in reverse as well. When white light (which contains all the colours of the rainbow) passes through an object like a star that contains different chemical elements, those colours distinctive to the elements are absorbed, leaving black gaps known as absorption lines in the spectrum. If, then, there is a black line in the yellow part of the light from a star corresponding to that key colour for sodium, you know that there is sodium present.

As far back as 1912, American astronomers Vesto Slipher and Milton Humason had noticed that some of the nebulae exhibited unexpected colours in the spectroscope. The pattern of lines was spaced out just as you might expect – but the colours were wrong. The lines had been shifted. We've all come across the way a different sort of spectrum can shift. When you hear the siren on an ambulance or police car coming towards you, passing you and moving away, the sound shifts in pitch. It's called the Doppler effect. What's happening is that the waves that make up the sound are being squashed together as the vehicle comes towards you and stretched out as it moves away, changing the pitch of the sound you hear.

We can't hear galaxies, but if they are moving towards us or away from us, the colour of the light undergoes a Doppler shift. If you think of light as a wave, then the Doppler effect makes the frequency of that light change – and we see the frequency of a light wave as its colour. If you prefer to think of light being made up of particles called photons, then the photons coming from a galaxy heading towards us have extra energy, while those from a galaxy heading away have less energy. We see the energy of photons as their colour.

Either way, when an object in space moves towards us its light becomes more blue – it undergoes a blue shift. As we've already discovered with the cosmic microwave background, if an object is moving away or the space in between expands, then its light shifts towards the red. There's a red shift. Slipher discovered such a red shift in a number of nebulae, which by the mid-1920s Hubble had identified as galaxies in their own right. By 1929, Hubble was able to report that almost every galaxy that had been observed was red-shifted. Apart from a few nearby exceptions, like the Andromeda galaxy which has a blue shift, all the galaxies are heading away from us.

What we are seeing here are two distinct effects. For (relatively) close structures like the Andromeda galaxy, there is a sizeable gravitational attraction between it and the Milky Way. The two galaxies are on a collision course. (There's nothing to worry about – we've about 4 billion years to go before they meet, by which time our Sun will probably have swallowed up the Earth anyway.) But most of the other galaxies are demonstrating something much more interesting. They are all red shifting, moving away from us. As we have seen, the universe as a whole is expanding.

It might seem strange that we appear to be at the centre of this expansion – because apart from a few anomalies like the Andromeda, every galaxy is shooting away from ours in all directions. But what is really happening is rather different. It's not so much that the galaxies themselves are moving, but that the space in which they sit is expanding. This is difficult to get your head around, because it is happening in three dimensions at once. It's easier to visualise in two.

Think of a balloon (you can try this at home). The surface of the balloon is effectively two-dimensional – it has no depth. Blow the balloon up a little bit and draw dots all over it. Then blow the balloon up some more. Each dot will stay at the same point on the balloon. They don't move on the balloon. They couldn't, they're fixed in place. But because the balloon *itself* gets bigger, the dots move away from each other. Pick any dot and all the other dots will move away from that point. Similarly, pick any galaxy in the expanding universe and all the other galaxies are moving away from it. Despite appearances, the Milky Way doesn't occupy a special position.

Even more extraordinarily, Hubble discovered that the further away a galaxy is, the faster it is moving. This effect was later named Hubble's law. Hubble himself seems not to have been particularly interested in the interpretation of the data, merely presenting it. But others such as the Russian physicist Alexander Friedmann showed how Einstein's general theory of relativity could describe an expanding universe, while the Belgian physicist and Roman Catholic priest Georges Lemaître specifically proposed a model of the universe where the universe had expanded, which would produce expansion rates that were strikingly similar to those that Hubble would publish.

Lemaître pointed out a fascinating consequence of this expansion, which you can visualise if you imagine making a video of the universe (or our balloon) expanding and then run it backwards. If we think of the balloon specifically, as we run the movie in reverse, the balloon gets smaller and smaller. If it were a very special balloon that could get as small as you like (rather than going all limp and floppy), then at a particular point in time it would vanish away to a single point. Now imagine doing the same thing with the whole universe. If you track it backwards, about 13.8 billion years in the past, the universe disappears to a point. This is around the point in time that is sometimes called the big bang.

Once we have set an age for the universe, we can say something about its size, at least in terms of setting a limit on how far we can see. If the universe formed 13.8 billion years ago, then the longest that light can have been travelling towards us is 13.8 billion years.* It can hardly have set off before the universe began. You might think this means that the visible universe should be 27.6 billion light years across (because light has been coming to us for 13.8 billion years in every direction), but that doesn't allow for the expansion we know has been taking place in the universe.

If light has been travelling 13.8 billion years to reach us, that light set off 13.8 billion years ago. But while it has been on its way, the universe has expanded. A lot. In fact, so much that most distant light shows us objects that are around 45.5 billion light years away. From this, we can say with confidence that the universe is at least 91 billion light years across. It may be bigger – it may even be infinite – but this is the limit of what we can see.

It is quite awe-inspiring just how far our picture of the universe has changed since the original ancient Greek idea that the solar system and a surrounding sphere of stars made up everything. All the way up to the Renaissance it was thought that there was nothing else, just the Sun, a few

---

* In practice, as we saw in Chapter 3, it's around 13.5 billion years, as the universe was not initially transparent.

planets and a hundred or so stars. With better equipment we thought that the universe was the Milky Way, which has turned out to have billions of stars in it. And now we know that there are other galaxies so big that they have as many as 100 trillion stars in them. Not just a few galaxies, either. There could be as many as 150 billion galaxies in the part of the universe we can see.

Many of those stars are very different from our own. But if we just take into account those that are relatively similar, there could be as many as 50 billion planets in the Milky Way alone. Until recently it has been impossible to tell if another star has planets. They are too far away to see in the reflected light of their star. But the motion of a planet around a star gives it a distinctive wobble.

By measuring the way that stars shimmy we can deduce just what is in orbit around them. At the time of writing nearly 4,000 planets have been discovered outside the solar system. It's easier to find the big ones which resemble our Jupiter (and which are unlikely to be habitable) because they have the biggest effect, but there are also some smaller planets already discovered that could support life. The universe is, without doubt, a remarkable place.

The back story of the universe

The big bang remains our best and most widely supported theory on how the universe began. It seems right, then, to start our exploration of dark energy and the expansion of the universe with a look at its lifecycle. As we do so, you should bear in mind the proviso that while the big bang theory fits all current observations, it had to be modified significantly to get that fit, and it is only one of several theories. For convenience I will describe the big bang as if it is fact, but (as is usual in science) it should be treated as the best available theory, not ultimate truth.

The big bang model begins around 13.8 billion years ago with the entire universe in the form of an infinitely small point called a singularity. A lot of people wonder 'What came before the big bang?', and in the basic big bang model the answer is very simple. Nothing came before the big bang.* Not only was there nothing before, there *was* no before, because our understanding of the big bang fits within a model of the universe that is based on Einstein's general relativity.

The general theory of relativity was one of the ultimate masterpieces of twentieth-century science. It describes how gravity acts. Back in the seventeenth century, Isaac Newton had described the basic law of gravitation, but he had not attempted to explain how gravity manages to pull things about from a distance. In the Latin original of his masterpiece the *Principia*, he wrote: 'Hypotheses non fingo' – I frame no hypotheses.

But general relativity does much more than account for gravity. It is a description of how space and time behave under the influence of mass – what stuff does to them. In this, it doesn't treat space and time as separate entities. Instead they are considered a single thing called spacetime, where time is another, rather special, dimension. And in the big bang model, spacetime begins pretty much at the big

---

* For the more pedantic, the big bang is strictly not the beginning of the universe, but the start of expansion an incredibly tiny fraction of a second later.

bang. The beginning of spacetime is the beginning of everything, including time itself. There can't be a 'before the big bang' because there was no time.

In this picture, science has no mechanism for explaining why the big bang took place. It just happened. You can say 'God did it', or 'It was spontaneous', or 'We don't understand why it happened' – the outcome is the same. In the standard big bang model, it is impossible to identify a reason, the universe just came into being as an infinitely small point. It is sometimes 'explained' as originating in quantum fluctuations in the initial vacuum – but this still begs the question of where the natural laws that make those quantum fluctuations possible came from.

In a tiny amount of time this entity began to grow. The very beginning, when the energy density involved is effectively infinite, is outside scientific theory, but once that expansion began, we have something we can apply science to. It's still a very small something, much smaller than an atom. It can seem puzzling that everything we now see in the universe – every star and planet, every bit of matter in every one of billions of galaxies – was compressed down to such a tiny size.

Admittedly, atoms are mostly empty space. By far the biggest component of your body is nothing. If you magnified an atom (any atom) until you could see the nucleus – the central part of the atom which has most of its mass – that nucleus would be about the size of a fly in an atom that was, relatively speaking, the size of a cathedral. Apart from a few electrons, zipping around the outside of the atom in a cloud of probability, the rest of it is emptiness. But even if there were some way in those first beginnings of the universe to do away with the empty space, we would still have a problem.

It has been estimated that if you took the entire human race and somehow removed all the space from their atoms (super-villains should take note – this isn't physically possible) they would be compressed down to about the size of a sugar cube. But we are talking about fitting what will become the entire universe – all those stars and galaxies – into a space much smaller than an atom. Taken at face value, this just doesn't seem right.

However, there are two important factors that make that cosmic seed possible. One is that the very early universe would not have had any matter in it whatsoever. As we have seen, Einstein's other great work, the special theory of relativity, included $E=mc^2$. This reflects the way that energy and matter are interchangeable. And energy doesn't take up space the way matter does.

The other contributory factor to squeezing the entire universe into a point is that, bizarrely, the universe really doesn't have to have started with much in it to end up the way it is. This is because gravity can be considered as a kind of negative energy.* Once you combine the mass of everything in the universe with the gravitational pull all that matter provides, they pretty well cancel each other out. There didn't have to be much content (perhaps not any at all) when the universe was first formed.

---

* Imagine two bodies starting a vast distance apart, not moving. There is no kinetic energy or energy due to gravity, as the attraction is negligible. But when they are closer, the bodies will accelerate towards each other, attracted by gravity. Both the gravitational and kinetic energy increase. Because energy cannot be created, this means that the gravitational energy must be negative compared with the kinetic energy.

The first significant event happened pretty early on. Around 10^{-36} seconds after the beginning, the universe is thought to have gone through a brief phase called inflation. This is expansion, just as we currently see the universe to be expanding, but it was on a phenomenally fast scale. A reminder that the minus sign before the little 36 means that 10^{-36} is the same as 1 divided by 10^{36}. And 10^{36} is 1 with 36 zeroes after it. So this inflation took place when the universe was just 1/1,000,000,000,000,000,000,000,000,000,000,000,000th of a second old.

Inflation wasn't a long process, but the growth was titanic. By the time the period of inflation finished, the universe was still only around 10^{-32} seconds old. This epic expansion took place in a ludicrously small fraction of a second. Yet in that time, the universe grew by at least a factor of 10^{30} and possibly as much as 10^{70}. Take those figures in slowly. The universe was at least 1 with 30 zeroes after it bigger than when it first started. We're still not talking about a huge universe – it was smaller than a present-day galaxy and quite possibly the size of a grapefruit – but it had been vastly smaller before. To expand so quickly, the universe would have to have stretched and enlarged far faster than the speed of light.

There seems to be a problem here, because Einstein's special relativity also tells us it's impossible for anything to travel through space faster than the speed of light in a vacuum (around 300,000 kilometres per second) – but we have to remember that when the universe is expanding, whether it's the 'normal' expansion that gives galaxies their red shift or the super-fast expansion of inflation, things aren't moving in space. It is space itself that is expanding, and there is no limit to the speed at which that can happen.

When the universe was just a fraction of a second old, it was so hot that the particles of matter that began to form from the pure energy were not atoms, nor even the familiar subatomic particles like protons, neutrons and electrons, but instead were quarks. We don't see individual quarks now because it requires a vast quantity of energy to break a proton or neutron apart, but in the sub-second universe these mysterious particles were the natural form for matter, forming what is known as a quark–gluon plasma.

By the time the universe was about a second old it had cooled enough (which basically means the individual particles had lost energy as it expanded) for protons and neutrons to form. But these were not just familiar matter – there was also an equal (or near-equal) amount of antimatter. Very soon most of the protons and neutrons combined with the antimatter equivalent, reverting to energy but also producing electrons and positrons (and other related particles). Within a few seconds these too had wiped each other out. Yet there seems an oddity in the symmetry between matter and antimatter that means that some of the matter particles would be left over. It's not clear if this effect is enough to explain the amount of matter we see – this is one of the weaker aspects of the big bang story – but we know that we don't now see lots of antimatter in the universe, so something must have happened to change the balance.

At this stage we're about three minutes into the life of the universe, and mostly it is made up of energy in the form of photons of light – but there is also matter, and for the next few minutes the universe has enough temperature and pressure to act like a massive star, converting hydrogen ions (a hydrogen ion is a hydrogen atom without an electron, which is just a single proton) into helium ions, just as the

Sun does today. This fusion process only lasted around a handful of minutes. By then the temperature had dropped from around 1 billion °C to a mere 10 million °C. The expansion of the universe had taken it beyond the limits required for the process that powers stars to continue.

By now the universe contained hydrogen, helium and a small amount of the next element up in weight, lithium. For the moment, all this matter was in the form of ions – positively charged particles, with electrons flashing around separately. The charge in this plasma made it difficult for photons of light to travel far, because they interact easily with charged particles, so this early universe of matter was opaque.

Following the end of the star-like period, a good amount of time elapsed as the universe expanded and cooled. After around 370,000 years those hydrogen, helium and lithium ions had lost enough energy to make it relatively easy for electrons to hook up with them, turning them from ions into atoms of hydrogen, helium and lithium. Now that proper atoms had formed, the universe became transparent. No longer were photons of light unable to travel more than a tiny distance before being captured. They began to flow freely, making this a very significant point in time, as we have already seen, by releasing what became the cosmic background radiation.

At this stage, thanks to that earlier period of incredibly quick inflation, the universe was very evenly spread with matter, with just tiny variations in density caused by the fluctuations that naturally occur at the level of individual particles, which obey quantum theory, predicting a randomness in behaviour. But over time, those tiny variations in the density of the universe began to grow. If all the matter in the universe had been spread out perfectly evenly, then the universe would have remained stable, but where there were little clusters of matter, other particles were attracted by the slightly larger gravitational pull.

More time passed. Between 100 million and a billion years into the life of the universe, these clumps of matter had got so big that they formed stars and the early beginnings of galaxies, structures that we now call quasars. This is a contraction of the term 'quasi-stellar objects', as quasars looked like extremely bright stars, but at their vast distance they are much brighter than any star could be. In a vast, evolutionary timescale the forms of galaxies and the stars we now know began to coalesce over billions of years.

Not all galaxies and stars formed at the same time. Although some stars in the outer halo of the Milky Way are over 13 billion years old, putting the galaxy's origins in the very early period of the universe, the main disc seems to have formed around 8.5 billion years ago when the universe was already over 5 billion years old. But the process of formation is a continuous one, driven by the insistent force of gravity. Our solar system, for example, began to come together around 5 billion years ago, forming from material from those earlier generations of stars as well as the remnants of the big bang that were spread throughout space. By 4.5 billion years ago the basic structure of the solar system we know today was in place.

Expansion goes on. The universe continues to cool as it has these 13.8 billion years. Stars still form and stars still explode. The evolution of the universe is a process that is still under way as we watch the skies around us. By the 1920s we had discovered that expansion, leading to the big bang model. That was remarkable enough in its own right – but it would be overshadowed by the discovery that the expansion was not constant, or perhaps slowing down due to the pull of gravity. It was accelerating.

GETTING BIGGER FASTER 5

Champagne supernova

One of Fritz Zwicky's important pieces of work, begun in the 1930s and continuing throughout his career, over and above identifying the action of dark matter, was research on supernovas. Stars go through an ageing process as they consume the hydrogen fuel available for fusion and go on to consume heavier elements. Some, like our Sun, are likely to fluff up into a red giant star, then produce a white dwarf in their old age. Others collapse dramatically as they reach the end of their lives. A particular type of white dwarf, of the right size (one larger than the Sun) will become so unstable in its collapse that it produces a vast nuclear explosion – a supernova.

There are several types of supernova depending on the size and constitution of the star that creates it. The earliest to be identified were type Ia and type II. Type Ia supernovas form when an old white dwarf star sucks in material from another star which cohabits a binary system with it. This pushes the white dwarf over the limit of around 1.4 solar masses up to which such stars are stable, producing a sudden, drastic nuclear reaction that blasts out much of the energy in the star. By contrast, type II supernovas occur when the core of a much bigger old, massive star, between around eight and 50 times the mass of the Sun, implodes. The outer part of the star is blasted outward in the visible part of the supernova, leaving behind one of two remarkable entities, a neutron star or a black hole.

As the name suggests, in a neutron star most of what is left after the implosion consists of neutrons. Without the repulsive force of the charged protons to keep them apart, the neutrons collapse into an incredibly dense mass. If you had a piece of neutron star about the size of a grape, it would weigh 100 million tonnes. What started off as a normal-sized star ends up on the same scale as the island of Manhattan.

If you were ever to come close to a neutron star it would not be a very comfortable experience. First of all, they are very hot. The surface of the Sun is about 5,500°C (the interior is much hotter, but it's the outside we experience directly). A neutron star, as the collapsed core of a star that went supernova, will typically have a surface temperature as high as 1,000,000°C. Also, because you can get much closer to a neutron star than to an ordinary star because of its lack of bulk, the tidal forces that the star produces in you would be much bigger than usual. Get close to a neutron star and there will be huge differences in attractive force between one end of your spaceship and the other.

The result, as the nearer end of the ship to the star is pulled away from the more distant end, is that your spaceship would be stretched out into a long, thin strip in a process graphically known to astronomers as 'spaghettification'. And the same process would go for you. You would be stretched into pink spaghetti. Neutron stars are the bad boys of the cosmic world. But even they seem timid when compared with the other potential product of a type II supernova, the black hole.

Black holes have become part of the mythology of space and fictional space travel. In movies they are often portrayed as totally black spheres in space with an irresistible pull, acting like vacuum cleaners that suck in everything around them. Get near a black hole and, Hollywood tells you, whatever you do you are inevitably going to be sucked in. The reality is rather different. A black hole has exactly the same gravitational pull it did before it collapsed – but you can get far closer to it even than is possible with a neutron star.

The gravitational forces that resulted in the formation of a black hole are so massive that they overcome the Pauli exclusion principle, the mechanism that usually prevents matter particles getting too close to one another. In principle, everything in the black hole has collapsed to a singularity – a dimensionless point similar to that posited for the big bang (see page 92). What we tend to think of as the outside of the black hole is actually its event horizon – the sphere at the distance from the black hole where gravity is so strong that nothing, not even light, can escape. The exact nature of black holes is extremely speculative as the numbers describing their behaviour reach infinity, which is an indicator that our theory has broken down. But all the evidence is that bodies approximating to black holes do really exist, both in this supernova-derived, star-sized form and in the supermassive black holes that appear to form the heart of most, if not all, galaxies.

These two types of supernova, Ia and II, whatever they leave behind, produce short-lasting, intensely bright torrents of light – sometimes brighter than the output of a whole galaxy. They are distinguishable from the Earth because the type II supernovas have hydrogen in their spectra and type I don't. (Type I supernovas are divided into three sub-categories, where a contains silicon, b contains helium and c contains neither.) As we have seen, astronomers use bright objects that tend to have a predictable level of brightness, such as the Cepheid variable stars, as standard candles to determine distances far beyond the capability of parallax. An ordinary star is not bright enough to be individually detected in distant galaxies – something much brighter, such as a supernova, was an obvious candidate for a new standard candle. What was obvious in theory proved hard to pin down in practice. It took around 50 years to the late 1980s before the suggestion was made that the type Ia supernovas would provide such a source of fixed brightness to deduce the distance to other galaxies.

This sounds simple enough, but in practice, supernovas have proved to be far less reliable standard candles than variable stars were. One problem was simply the difficulty of spotting the flare-ups. Supernovas are individual events that are often visible from Earth for only a few days or weeks, and finding them required painstaking comparisons of the views from telescopes taken at different times in the hope that what seemed to be a new star would have appeared. Early searches went months at a time without seeing a single new supernova.

The other big problem in using supernovas as standard candles proved to be consistency. By the early 1990s it had become clear that there was considerable variation even within the type which seemed the best bet as a standard candle, type Ia. This was made obvious when two type Ia supernovas were spotted in the same galaxy – so they were at approximately the same distance from the Earth – but one was ten times brighter than the other. Even if two supernovas did happen to have the same brightness, they could appear different at the same distance if one had more interstellar dust in the way than the other – far more of a problem than with variable stars, which were far closer than the distant galaxies being studied. The whole exercise of using supernovas as standard candles was fraught with potential difficulties.

Thankfully, new technology was able to provide astronomers with a lifeline. From its earliest days, astronomy had relied on the ability of the human eye to detect light from the sky, enhanced since Galileo's time by lenses. The next big step forward had been the addition in the 1880s of photographic plates to the astronomer's armoury, which could collect light from the same source over a period of time (provided the telescope was constantly moved to account for the rotation of the Earth). This effectively amplified the light from the star. But from the 1980s onwards, increasingly sophisticated electronic detectors – cousins of the cameras used in modern phones – could be deployed to collect incoming photons and build up a more sophisticated picture of the sky over time.

The advantage that these detectors had over photography was that they made it much easier not just to build up a more intense image but to be able to monitor the output of a cosmic source as time ticked by. This was a crucial benefit, as supernovas go through different stages during the stellar explosion, first brightening then dimming. By plotting out the changes in brightness, showing the rate at which the brightness varies, a chart known as a light curve can be produced. Using these, it proved possible to refine the distinctions between supernova types far more than dividing them into types – each effectively had its own 'fingerprint' in its light curve. And this brought type Ia supernovas back as potentially useful standard candles. But once this technique existed, it was no longer necessary to try to use them as standard candles: now they were *calibrated* candles. Given the shape of a light curve, it was possible to deduce the distance to the supernova.

Supernovas would play a crucial role in the discovery of dark energy, along with that other astronomical favourite, red shift. We've already seen that Hubble and others used red shift to measure the rate at which the universe is expanding, by looking at the red shift of different galaxies. Combining the red shift of galaxies and the distance to those galaxies, astronomers hoped to be able to detect the rate at which the expansion of the universe was slowing down.

The idea that the expansion had to be slowing was pretty much a given. Everything in the universe attracts everything else – gravity has no limits. So, over time, it was to be expected that gravitational attraction between galaxies would gradually overwhelm the expansion of the universe. By observing the red shift of very distant universes – seeing them as they were billions of years earlier because of the time light took to arrive – and combining that with the distance to them, it should be possible to work out how much the rate of expansion was slowing down from those early years to the present.

Astronomers vs. physicists

As the understanding of supernova types – and the technology to detect them – became more advanced in the 1990s, two teams went head to head in the attempt to measure the deceleration of the expansion of the universe. They came from very different backgrounds. As might be expected, one team were astronomers – but the others were physicists.

Astronomy predates physics by many hundreds of years, and astronomers have always tended to do things their own way. This division arguably dates back to the historical division between mathematics and natural philosophy – the latter now called science. Although we now put mathematics and science on a continuum, to the extent that, for example, the former Cambridge workplace of Stephen Hawking was the Department of Applied Mathematics and Theoretical Physics, originally astronomy was the only numerically-based science, and so was considered part of mathematics, whereas the sciences were qualitative.

Now, astronomy tends to be lumped in with physics, but there is often a different mindset between the practitioners of the disciplines. When it came to working on supernovas, a group based at one of the most active physics hubs in the United States, the Lawrence Berkeley Laboratory, felt that they had insights that would give them the edge over their astronomer competitors.

The group was part of the Center for Particle Astrophysics at Berkeley, which we have already met in the field of dark matter detection. Dark matter was, indeed, the main focus of this institution, established in 1988. Staff at the organisation were trying to detect dark matter particles, both directly and in the impact dark matter had on the cosmic microwave background radiation. But those running the Center felt that it was also important to have a more accurate idea of just how much matter there was in the universe – both dark and ordinary. And that would involve using the red shift of distant galaxies to calculate the deceleration of the expansion of the universe, which in turn would provide an estimate of how much mass of matter would be required to cause such a braking effect.

Although the Berkeley team were physicists, some already had experience with the search for supernovas. Saul Perlmutter and Carl Pennypacker had spent the mid-1980s working on automated systems to help pick out supernovas from other bright lights of the night sky. Historically, this had been done by taking a pair of photographic plates of the same region of sky, then repeatedly flipping between the two of them in the hope of spotting a difference by eye. But electronic scanning had made it possible to take the data from two images, subtract one from the other and look out for contrasting spots that remained – these would be objects that had appeared or disappeared between the images being taken, the hallmark of a supernova.

Along with Pennypacker's supervisor, Richard Muller, and a few graduate students, Perlmutter and Pennypacker had some success in supernova hunting, using a 760-millimetre telescope at the Leuschner Observatory in Lafayette, California. The first detection by the so-called Berkeley Automatic Supernova Search (BASS) team was in May 1986. But they only managed to produce data for a handful of nearby supernovas, which didn't give the requisite leverage to make calculations on the contents of a wider universe.

To get a useful picture of the deceleration of the expansion would mean getting data on much more distant galaxies. The team needed bigger guns than the relatively small telescope they had been using and were able to get funding to produce a new, much higher-specification camera which was to be mounted on the impressively large

3.9-metre* Anglo-Australian telescope which dated back to the mid-1970s and was based at Siding Spring Observatory in New South Wales, Australia.

The downside of using such a telescope that was (and still is) in high demand was that the observing time allocated to the team was strictly limited. (Those allocating time might even have been particularly stingy as the Berkeley team weren't 'real' astronomers.) Over the period of around 30 months in the late 1980s that the project was run, the team was allocated a total of twelve observing nights. Of these, only two and a half were usable. Initially six candidate observations of supernovas were made, but on further analysis all six were eliminated. In trying to peer further into the depths of the universe than had ever been possible before, the team were working at the very limits of the available technology and after three years of work the result was a blank.

This kind of noble failure is relatively common in science. Many experiments simply don't work out. In principle this is a good thing. If scientists only ever took on projects that were certain to succeed, they would not be pushing the limits of knowledge. It's only by taking risks that we progress. However, scientists are also human – and it would be unfair to suggest that this was not a painful setback for those involved. It didn't help that the Berkeley team now had to justify repeating what had been a failure to those in charge of the purse strings. However, they managed to persuade the bureaucracy at the Center for Particle Astrophysics that

---

* For those more familiar with the inch-based measurement that used to be common on, say, the 100-inch Mount Wilson or 200-inch Mount Palomar telescopes, the 3.9-metre Anglo-Australian telescope has a 154-inch mirror.

the problem had not been down to their methodology. They needed a still better camera and a telescope with a more favourable location.

At first sight, the choice of instrument that the Berkeley group proposed was strange. They wanted to move to the Isaac Newton telescope. This 2.5-metre telescope had seen first light in 1967 when it was constructed at the Royal Greenwich Observatory. Despite the name, the observatory is not located in the London borough of Greenwich, but rather in the grand surroundings of Herstmonceux Castle in Sussex, where the observatory was moved in the 1950s to improve viewing conditions. Similarly, the telescope itself was moved in 1979 to La Palma in the Canary Islands, off the coast of North Africa, where the weather was far better for observing than in cloudy England.

The Berkeley team hoped that the climate in the Canaries would allow significantly more observing time than had been possible at the Australian site. Combined with an enhanced camera this should more than make up for the somewhat smaller telescope. Certainly, there wouldn't be another chance for the team. If this attempt failed entirely like its predecessor, the game would be up.

Supernova archaeology

The first breakthrough for what was by then known as the Berkeley Supernova Cosmology Project (or SCP) came in May 1992, when comparisons were being made between recent images and ones taken in March of that year. The requirement was to spot a change, though even these could be eliminated if the new arrival turned out to be a local object such as an asteroid that had strayed into the field of view. However, in May, Saul Perlmutter found a change in intensity at one spot that could not be ruled out.

Over the next few weeks he pestered astronomers around the world to check out his potential supernova. One sighting wasn't enough – it needed confirmation and more data than was possible from the SCP sighting to be worthwhile. Telescope time is strictly limited and is usually booked up months in advance for specific projects – Perlmutter was asking astronomers to briefly put their own work aside to try to confirm his finding, and he wasn't even part of the community himself. Yet he managed to persuade enough others to provide supporting data to be able to confirm the existence of the supernova and plot out its light curve.

Things went less smoothly when it came to detecting the red shift of the supernova, which would be crucial in establishing how the galaxy was moving with respect to ours. Remarkably, Perlmutter persuaded yet more observatories to take the spectroscopic reading necessary to pin down the red shift on twelve different occasions. Remember, deducing the red shift depended on seeing how much the collection of lines produced by various elements were shifted from their actual position in the spectrum from the supernova. Of the twelve observations that Perlmutter requested, eleven proved to have bad weather, and on the twelfth, the technology failed. There was still no spectroscopic data for the new supernova.

By now it was August 1992. Perlmutter had a last shot at persuading a very reluctant astronomer to take a reading. The 4.2-metre William Herschel telescope at the same observatory site as the Isaac Newton on La Palma managed

to produce usable spectra. The SCP team had fully captured a first distant supernova, which would prove to be
record-breaking in its red shift. The most distant previously
discovered supernova had looked back around 3.5 billion
years in time – this one reached back 4.7 billion years,
around one-third of the lifetime of the universe.

This discovery was treated with some suspicion by the
astronomical community, particularly those specialising in
supernovas, who felt that Perlmutter and his colleagues
did not have a good enough grasp of the practicalities of
observing, nor the right skills for dealing with factors such
as intervening dust. But by 1994, the Berkeley SCP team had
taken a further step forward. In the early months of that year,
they managed to discover six distant supernovas in just six
nights of observing.

The astronomers strike back

The striking success of the Berkeley team, who, compared with previous attempts, now seemed to have a
production-line approach to distant supernovas, spurred a
group of astronomers including Brian Schmidt from Harvard
University and Nicholas Suntzeff, who was working at an
observatory in the clear skies of Chile, to take on the inter-
loper physicists.

Scrambling to produce the essential software that was
needed to compare computer images from observations,
these rivals were able to make their first observations from
the Chilean observatory (complicated at the time by Schmidt
having a posting in Australia). Like the Berkeley team, most
of their attempts failed but a final observation squeezed out

a new supernova, this time 4.9 billion years back, beating the Berkeley record.

The SCP team, meanwhile, had not been standing still. Through 1994 and into 1995 they had developed an industrial-scale approach to supernova discovery. They would observe hundreds of galaxies in one night, then several weeks later would observe the same set. With more advanced software they were able to detect potential supernovas within hours and could then get other observatories on track to confirm these sightings and take spectroscopic readings. Now that the SCP's scientists were no longer inexperienced interlopers, they could book this telescope time to back up their observations formally, rather than scrape around for support as they had been forced to do previously.

By now, then, the Berkeley SCP physicists had a whole string of observations, while the Harvard and Chile astronomers had one. To an extent, this reflected a very different approach. The physicists' take on the process was 'find supernovas first and worry about what to make of them later'. The astronomers, by contrast, were less concerned about finding too many sightings initially, wanting instead to focus on pinning down the interpretation of the data. Were these type Ia supernovas at all? Could intervening dust be distorting the readings – and if so, what could they do about it? And how was it possible to use these supernovas to fix distance effectively?

It was only in the autumn of 1995 that the career astronomers, by then known as the High-z group ('z' is the symbol used for red shift – the higher the z, the faster the galaxy is receding), was ready to follow up their initial discovery with a series of observations in Chile. They were confident now that they could accurately use type Ia supernova light curves and red shift data to make observations about the rate of expansion of the universe. To add immediacy to the competition between the two teams, they were now both using the same telescope in Chile on alternate nights. As a result, the SCP team had brought their total observations up to 22 supernovas.

The Berkeley physicists decided to go for broke in the race and asked for time on the Hubble Space Telescope. Although smaller than many of its terrestrial equivalents at just 2.4 metres across, the satellite-based telescope had the potential for unrivalled observation of extremely distant supernovas because of the lack of atmospheric interference. Launched in 1990 and made serviceable by a Shuttle mission in 1993, the Hubble provided the definitive opportunity for trouble-free supernova searching. There were no bad weather nights in space.

Hubble's director, Bob Williams, was inclined to give the Berkeley physicists time, but consulted the High-z team. Their initial inclination was to try to discourage Williams from allowing their rivals access to the telescope, pointing out that the Hubble telescope was only supposed to be used on projects that wouldn't work from Earth. However, Williams felt that the decision lay within his discretion. At the last moment, the High-z team realised they were in danger of preventing themselves from using the best possible tool for the job simply to spite their rivals. Williams duly granted both teams access to the space telescope.

It took another two years through to September 1997 before both teams had got sufficient data to start to work out the rate of slowing of the expansion of the universe. However, before long it became clear in both camps that something wasn't quite right. Looking far out into the depths of the universe, red-shifted supernovas were less bright than the expected distance for that red shift predicted they should be. The supernovas were further away than they were thought to be. It seemed that something – some unknown source of energy – was causing the rate of expansion of the universe to increase rather than slow down. The stage was set for a second dark contributor to the universe.

The day cosmology became a science (or not)

As we saw in Chapter 2, cosmology, the study of the universe as a whole, had a history of being more speculation than any other part of science. It began as religion or pure philosophy, with models of the origin of the universe plucked out of the air. Whether the universe emerged from flames or an egg (literal or metaphorical), there had been no data, no way to test these theories by experiment or observation. In such a situation, it was arguable that what was being dealt with was speculative fiction rather than true science.

However, at a press event on 8 January 1998, representatives of both the SCP and High-z groups, along with two other teams which had been working on the output of distant galaxies with high radio energy signals and on the large-scale structures of the universe and their relationship to the cosmic microwave background, put forward what they believed was definitive data that took predictions of the future of the universe from speculation into the realm of real science.

As an aside, this is, strictly speaking, an exaggeration. For example, the big bang model has had to be modified several times to match observation. In principle, practically any theory can be patched up until it appears to work this way.

It's rather like the way that Aristotle's model of the universe had been patched up with epicycles. As the patching is based on matching what is observed, it inevitably works, but that doesn't make it right.

While the big bang with inflation remains our current best accepted theory (just as the existence of dark matter is our current best accepted theory for the odd gravitational behaviour of galaxies and galactic clusters), a good number of physicists have doubt about its foundations. Inflation, for example, is seen by some to lack justification. And the origin of the universe from a singularity is widely considered to be an indication of theory breakdown, as it requires infinite values.

However, even though the combination of big bang and inflation *could* still be incorrect, there is no doubt that the January 1998 announcement saw our picture of the way that the history of the universe is unfolding go from pure speculation to speculation that had a sufficiently strong match to observation to make it feel like science. Saul Perlmutter made this clear at the event, commenting: 'For the first time we're going to actually have data, so that you will go to an experimentalist to find out what the cosmology of the universe is, not a philosopher.'*

The data appeared to show not just that the universe was never going to stop expanding but that something extra – dark energy – was accelerating that expansion. Strangely, such a factor had appeared early on in the theoretical world of the general theory of relativity as a fudge that Einstein had inserted into his equations to counter the fact that, as they

---

* It's not recorded if anyone pointed out to Perlmutter that 'cosmology of the universe' is tautological.

first were formulated, the universe was unstable. This factor is known as the cosmological constant and is represented by the capital Greek letter lambda (Λ).

Cosmological constant

To see the relevance of the constant, we need to dip our toe in the general theory of relativity. As we have seen, this was Einstein's masterpiece, put together in the years leading up to 1915, and describing the action of gravity as a warping of space and time caused by matter. The field equations of general relativity can be simplified in the relatively friendly form:

$$G_{\mu\nu} + \Lambda g_{\mu\nu} = (8\pi G/c^4)\ T_{\mu\nu}$$

... where, broadly, the part on the left describes the curvature of spacetime and the part on the right the mass-energy causing it. (The equations are a lot more complex than they look, as each part with subscripts is a tensor, a multidimensional geometric object which, in this case, corresponds to ten different equations.) Originally Einstein formulated the equations without the lambda part, but added it in to account for what he assumed to be an error. In 1917 he applied his theory to the universe as a whole. His equations seemed to predict a universe that was going to expand for ever or collapse under the attraction of gravity. But Einstein was convinced that neither of these options was the case – his constant was introduced to counteract the effect of gravity and leave the universe static.

It didn't work out very well. The slightest shift from that equilibrium and the universe would begin to contract or expand. Later, Einstein would call lambda his 'greatest mistake', as the observed expansion of the universe (assumed to be slowing due to the impact of gravity after the dramatic first burst of inflation) did away with the need for it – in effect, for many years lambda was assumed to be zero. What the 1998 announcement showed was that to accurately reflect what was happening in the universe, the cosmological constant would have to have a greater effect than Einstein ever assumed. It would have to be big enough to drive an accelerating expansion. The value for the cosmological constant announced in 1998 remains approximately the accepted value of around 10^{-52} m^{-2}.*

Although Perlmutter proudly announced the movement of cosmology into the field of science, this value proved to be a serious problem for physics, or at least for the version of physics that arises from our current models. And though this problem has now been around for over 20 years, it still presents exactly the same nightmarish issues. To understand it, we need a brief excursion into the quantum world.

The quantum void is not empty

One of the fundamentals of quantum physics, the physics of the very small, is the uncertainty principle, introduced by German physicist Werner Heisenberg in 1927. This says that there are certain parameters of reality that at the quantum level are inextricably linked. They vary in such a way that,

---

* If something grows by 1 metre in the first second, 2 metres in the second second, 3 metres in the third second and so on, the acceleration is 1 m^{-2}. The scale of the cosmological constant is 10,000 trillion, trillion, trillion, trillion times smaller.

should we know one of these parameters in detail, we will be less and less certain about the value of the other. The best known of these pairings is between momentum (mass times velocity) and position. The more we know about the momentum of a quantum particle, for example, the less we know about its position, and vice versa. But arguably the most dramatic of these pairings is that of energy and time.

What this uncertainty implies is that if we observe a volume of empty space, and pin down that observation to a very small timescale, the energy present in that volume can be anything from very small to very large. Sometimes it will be so large that it is enough to produce matter. This means that quantum physics suggests that empty space should be a seething mass of so-called 'virtual' particles that pop into existence and disappear again before we can directly detect them.

There is good evidence that this really happens in something called the Casimir effect, named after Dutch physicist Hendrik Casimir. He predicted, and experiments have repeatedly shown that, for example, if there are two flat sheets of metal placed very close together, then they will be attracted towards each other. This is effectively because there are far more virtual particles popping into existence outside the sheets than is possible in the extremely narrow gap between them, and when these particles collide with the sheets before disappearing they give the sheets an inward push.

So empty space was predicted to contain a kind of energy – the energy left when this variable energy, sometimes called vacuum energy, is averaged out. It seemed reasonable that this would be the energy that was driving the cosmological constant – what we now call dark energy. But when quantum field theory is used to predict what the cosmological constant should be if it *were* powered by vacuum energy, it comes up with a value of around 10^{68} m^{-2}. This means that theory and the observation differ by a factor of 10^{120}. This is by far the biggest deviation between theory and practice in all of science. It is *so* wrong that there is clearly a huge incorrect assumption somewhere – we just don't know where.

Pinning down the future of the universe

As data began to be assembled, it initially appeared reasonable that lambda was zero. However, we need to bear in mind the huge uncertainty in those early measurements which had come together between 1995 and 1997. With factors such as dust and the limitations of measurement playing a part, while this assumed situation could fit with the data, so could a wide range of other possibilities. But new techniques, using a range of filters to attempt to separate off the reddening effect of dust, were coming into use. And as data continued to pour in from more and more supernova observations, at increasingly deep distances in time, it seemed that the early assumptions were incorrect.

When the more accurate data from the Hubble Space Telescope was added into the mix, the distribution shifted. The supernovas were fainter than might be expected for the degree of red shift that they had. But as yet there were relatively few data points. The Berkeley team were reluctant to publish contradictory information, but in the end did so because the approach, using Hubble, was different. Their new data (with plenty of 'if's and 'but's) was published in the journal *Nature* in October 1997. The same month, the High-z team released its own results from Hubble. They too contradicted the conventional view of a simple universe with no cosmological constant and just the right amount of mass to explain what was happening.

The agonising the teams went through at the end of 1997 arguably demonstrates that, despite the claims, cosmology was still some distance from being a true science. Many of those involved did not want there to be a cosmological constant and were searching for ways to do away with it. This comes across clearly in an email exchange between members of the High-z group about whether to go public with their Hubble Space Telescope results. At one point, Brian Schmidt, arguing for publication, comments: 'As uncomfortable as I am with the Cosmological constant, I do not believe that we should sit on our results until we can find a reason for them being wrong (that too is not a correct way to do science).'

By February 1998, the matter was coming to a head. Both teams were attending a conference (ironically, a meeting that was primarily focused on dark matter) at the University of California, Los Angeles. The more conservative members of the High-z team were still inclined not to draw any significant conclusions. Like Hubble with the original red shift data, they wanted to just present the data and let others draw the conclusions. Yet the press had been picking up on the SCP group's more impressive-looking collection of supernovas. Some in the High-z group were convinced that their Berkeley rivals were about to go public on the existence of a cosmological constant driving the acceleration of the expansion of the universe. High-z were due to formally publish on the subject a couple of weeks after the conference, but the feeling was that this was too good an opportunity to miss.

After sitting through Saul Perlmutter presenting the SCP data and only going as far as to say that there was some evidence for the existence of a cosmological constant, the High-z spokesperson Alex Filippenko took the plunge and announced that their data showed that the constant was real.

The response to this announcement was mixed. A surprising number of astrophysicists went along with the cosmological constant, even though the value was a ridiculous factor of 10^{120} away from the predictions of quantum theory (see page 118). One of the reasons seemed to be that the existence of a driving energy for expansion supported the most hypothetical aspect of the current big bang model, the idea that the universe inflated extremely quickly in the first fraction of a second. And the data being presented suggested that the universe had to be at least 15 billion years old (a figure later refined to 13.8 billion), which gave time for the large-scale structures that astronomers had observed to have formed.

Others, however, felt that there had to be an alternative explanation. They did not doubt the data – it had come from two separate teams, working in opposition and using different approaches – yet all data is subject to interpretation, and the indirect nature of cosmological data made interpretation particularly open to misleading results. Inevitably there were big assumptions being made – ones that were difficult, if not impossible, to test. Mostly these assumptions were about consistency.

Is the universe consistent?

Remember that an absolutely crucial foundation of these observations was the use of type Ia supernovas as standard candles. The early crises of supernova research had shown that type Ia supernovas were not standard candles in the same sense as, say, variable stars, which had a brightness that corresponded directly to the timing of their variation. But the supernovas had been tamed by using the curves of their changing brightness to produce a way of calibrating their distance.

This calibration depended on observations of relatively near supernovas, where it was possible to confirm the distances using other types of standard candle. But to get the cosmological constant results, it had been assumed that the same calibration could also be applied to supernovas in very distant galaxies. Bear in mind that the further out you look into space, the further back you look in time, because light takes time to reach us. The distant galaxies were being observed as they were billions of years ago. But what if some factor had been changing with time, so that the output curves of these far older supernovas were different from the more modern ones? In that case, the assumptions about how far away the galaxies were would be completely wrong.

Similarly, consistency assumptions were being made about the impact of interstellar dust on observations. A better understanding of the impact of dust (which inevitably makes a distant supernova look dimmer than it actually is) had been one of the reasons the High-z group were able to catch up with the impressive lead that the SCP physicists had. Dust is a tricky business, and continues to be to this day, as another group of experimenters discovered to their cost.

In 2014, an experiment based at the South Pole called BICEP2 (the BICEP part stands for Background Imaging of Cosmic Extragalactic Polarisation) reported that it had produced data that showed evidence of gravitational waves from the very early moments of the universe. As we have seen, astronomers have a real problem looking back to before the universe was around 370,000 years old, as before that point all the different forms of light used in astronomy, from radio to gamma rays, were unable to pass through the universe. However, nothing known can block gravitational waves.

If inflation had indeed happened shortly after the universe came into being, it was expected that this violent expansion of spacetime would have set up so-called primordial gravitational waves, which would still be travelling through the universe. This was what BICEP2 was reported to have discovered, giving a rare piece of supporting evidence for the existence of inflation. However, within months, doubts were being raised about the results. It is now accepted that the apparent signal came from dust.* When the impact of dust was properly taken into account, there was nothing left.

In the case of the cosmological constant findings, it had been necessary to remove the impact of dust by deducing the reddish spectrum imposed by the dust and subtracting it from the data. As with the supernova observations, the assumption was that the impact of relatively close dust, which was far easier to study, would be the same as dust everywhere in space, including in those distant, early universe views. But what if there were so-called grey dust

---

* The BICEP2 experiment was looking for polarisation in the cosmic microwave background (polarisation is when light is only vibrating in some of the possible directions at right angles to its direction of travel). The hope was that polarisation would have been caused by the primordial gravitational waves. However, when light reflects off objects it is often polarised – this is why polarising sunglasses reduce reflected glare. And when the radiation was scattered by interstellar dust – reflecting off individual dust particles – it was also likely to become polarised. The measured polarisation proved to be exactly as expected from dust scattering.

between the early galaxies, which had a different optical impact? This could explain away the unexpectedly dim light from supernovas in distant galaxies and do away with the concept of dark energy.

The prehistoric universe

With the data that existed at the time of the press conference in 1998, there was no way to eliminate the possible factors that could be distorting the data. Yet theory suggested that there was a way to see exactly what was causing the unexpected dimness. At that time, the most distant galaxies that had been studied were being seen as they were around 5 billion years ago. But if it were possible to see objects with a much higher red shift – hence much further back in time – it was expected that things would be different. The early universe was much smaller, making the density of matter much higher. So high that gravitational attraction should have been beating the expansionary pressure of dark energy.

This would mean that back then, the expansion of the universe should have been decelerating. It was only after the contents of the cosmos had thinned out sufficiently for dark energy to dominate that the acceleration should have kicked in. If supernovas in distant enough galaxies could be observed, then this should provide a way to distinguish between dark energy and the other potential causes for the observations.

Luckily, the Hubble Space Telescope had already been used to make images that could provide exactly the observations that were required – the study known as the Hubble

Deep Field. This was a set of observations made in 1995 of a very small area of the sky (similar in area to looking at a tennis ball from 100 metres). A comparison had been made with the same stars in 1997, and a pair of very distant supernovas detected, though this data was not accompanied with the associated spectral and light curve analyses to deduce the distance and rate of motion. Yet in 2001, the Deep Field data would prove to be remarkably valuable.

The breakthrough was the brainchild of Adam Riess, a former member of the High-z team who had come up with the light curve analysis of supernovas, and who would share the Nobel Prize in Physics for 2011 with Saul Perlmutter and Brian Schmidt for their work on dark energy. Riess was aware of the two supernovas in the Hubble Deep Field, and in 2001 decided to spend a little time on an extreme long shot. It was very unlikely to pay off, but if it did, the reward would be remarkable.

Nobody had intentionally made the necessary measurements on the two supernovas back in 1997. But what if someone had just happened to collect that data as part of another study at the time? It would require a considerable fluke, but it might have happened. By coincidence, a new instrument had been added to the Hubble telescope in 1997, which included a spectrometer. And that technology was being tested at just the right time. Serendipitously it caught one of the extremely distant supernovas from the 1997 Deep Field data. Riess was able to work out the red shift, putting the view at over 10 billion years in the past – far enough back that, according to theory, dark energy was yet to have dominated. And yet this supernova was around twice as bright as it should have been if either of the alternative causes (changed light curves or different dust) held true.

At the time of writing, the most distant type Ia super-nova to be identified is SN UDS10Wil, discovered in the Hubble's CANDELS Ultra Deep Survey, which gives a view back of around 10.5 billion years and also supports the earlier findings. Dark energy, it seems, is here to stay.

Cosmological constant or quintessence?

Interestingly, the aspect of consistency would continue to haunt dark energy research in a different way. To come up with a value for the cosmological constant it was assumed that dark energy had always been on the same scale, and was the same everywhere throughout the universe. It's interesting that dark energy scientists seem to have been rather more open in this respect than those interested in dark matter. Most physicists assumed the consistency of gravity and were unhappy with modified Newtonian dynamics approaches. But far more, perhaps because of the initial resistance to the existence of the cosmological constant, were happy to consider dark energy not to be a constant at all but a field – a value that varied from place to place in space and time.

The theorists had even given this hypothetical cosmological field a cute name – the quintessence. This was a name dating back to ancient Greece. As we have seen, for well over a millennium and a half it was accepted that Earthly substances were made up of four elements – earth, water, air and fire. But this was only considered to be the makeup of the region of the universe below the orbit of the Moon. From the Moon upwards, everything was thought to be made of a different fifth element or essence – the quintessence.

It's arguable, though, that whoever was personally responsible for the new use of this term* didn't know his history of science. The ancient Greeks were aware that substances on the Earth changed with time and location. That was the reason everything from the Moon outwards could not have been made of the standard elements. The heavens were considered to be immutable – the essence of the quintessence was that it never changed. Ironically, the modern cosmological quintessence, if it exists, is defined by being something that *does* change – otherwise it would just be the plain old cosmological constant.

Since Riess's breakthrough announcement, there have been a range of measures used to attempt to get a better grip on the progress of the expansion of the universe, and hence the nature of dark energy, over time. As well as taking many more supernova observations, astronomers have made use of the gravitational lensing effect (see page 5) of very distant galactic clusters to try to get a feel for the impact of the expansion of the universe, have hunted for vast structures that might reflect waves in the early universe, and have searched the cosmic microwave background radiation for photons that had received a boost of energy when passing through a galactic cluster, all frustratingly indirect methods.

Most evidence to date seems to suggest that a cosmological constant is more likely than a variable quintessence. For example, in 2016 data combined from NASA's Chandra X-Ray observatory and the Hubble Space Telescope was used to study galactic clusters ranging in time from 760 million years in the past to 8.7 billion years. The study showed no

---

* It first appeared in a 1998 paper by Robert Caldwell, Rahul Dave and Paul Steinhardt.

variation in the scale of dark energy in that period (though, of course, as the impact of gravity became less due to the universe expanding, dark energy became dominant in that timeframe).

A more important input from the cosmic microwave background is that the background shows that the universe is pretty much flat. This is in a geometric sense. In principle, space could be curved in such a way that it turns in on itself, or it could curve outwards rather like a saddle in shape. In practice all the evidence is that it is pretty much flat, which is backed up by the relative uniformity of the background radiation. When we look at the satellite plots of the cosmic microwave background it appears to vary considerably, but this is a result of the way the data is presented. From the darkest to the lightest part of the plot only reflects a tiny fractional variation.

The way that the variations are distributed in the background radiation, particularly using a measure known as the angular power spectrum, emphasises both the flatness and that around 95 per cent of the content of the early universe seems not to interact electromagnetically. This flatness of the universe's spacetime structure also comes through in the distribution of distant supernovas used to discover dark energy. Theoretical models based on general relativity give a figure for the average density of matter in the universe that should correspond to the universe being flat. With just the estimated amounts of ordinary and dark matter present, the density comes out as far too low for the universe to be flat. But add in the dark energy and the outcome is a density that pretty much exactly corresponds to flatness.

The fact remains that we have no standout theory for what dark energy actually *is*. 'Dark energy' is just a name – it tells us no more than calling it 'fluffy bunnikins'. It's not that there haven't been theories as to what is causing the acceleration of the expansion of the universe. In a talk at a 2007 conference, Brian Schmidt listed around fifty different such theories, with names ranging from the brand-name like 'radion' and 'Dilaton', through technical sounding concepts such as 'scalar + spinor' and 'adiabatic matter creation', to the downright bizarre-sounding 'Pseudo-Nambu-Goldstone Boson Quintessence', and 'k-chameleon'.

For the future, different avenues are being suggested to find out more about the nature of dark energy. Supernova measurements continue to be a favourite tool, and gravitational lensing by very distant objects is also liable to have a role. A relatively new option is the study of baryon acoustic oscillations.

When what is now the cosmic microwave background first set out as high-energy gamma rays, the electrically charged plasma that still existed widely in the universe would have undergone a set of physical vibrations. Because these were vibrations in matter, they were similar to sound waves – so-called 'acoustic oscillations'. These waves travelling through the matter are likely to have influenced its distribution to eventually form galaxies. The suggestion is that these waves should have resulted in galaxies forming preferentially with gaps of around 490 million light years between them.

In principle, this gives us a new kind of standard candle for reaching back to these early days of the universe. If you know two newly formed galaxies were actually these kinds of distance apart, given their apparent separation from each other as seen from Earth, you can work out how far away they are from us. Like all approaches based on standard candles, there is an element of uncertainty here. Yet it demonstrates how the search for mechanisms to probe the influence and causes of dark energy is continuing.

Imaginary darkness

Like dark matter, there is a possibility that dark energy is an error introduced by assumptions. A fundamental assumption for cosmological models is that the universe is pretty much the same in all directions and that there is nothing special about our location. However, it is possible that our galaxy is not in a typical part of the universe, but rather one with less matter in it than is generally the case, which could give a misleading picture of what is happening elsewhere. Or it could be just that we simply don't have enough data yet to get a clear picture of what is happening. Even the hundreds of supernovas now captured provide a tiny sample from the whole universe.

However, there are few cosmologists or astrophysicists who doubt the reality of dark energy. Unlike dark matter, there isn't a good alternative theory that explains some aspects of what is observed better than the currently accepted one. Questions remain about the assumptions being made, but that gives no guidance for alternatives.

Until and unless we get significantly updated data, dark energy is here to stay.

A CONTINUING STORY 6

The search continues

Despite failure to date, existing dark matter detectors are continuing to be used and we can expect a number of other possibilities to open up to, at the very least, help to rule out some options.

Next-generation direct detectors with greater sensitivity and (hopefully) ability to distinguish dark matter from spurious readings are being constructed. The PandaX detector in China (see page 57) is being upgraded with a much bigger xenon target: this 'PandaX-xt' is expected to go live in 2020. Similarly, in Lead, South Dakota, the xenon-based American LUX detector, which was active between 2013 and 2016 (finding nothing) is being upgraded as LUX-ZEPLIN, with over twice as much xenon as the new PandaX design. This also should be ready by 2020.

Indirect observation is also being stepped up. The US-funded Large Synoptic Survey telescope under construction at Cerro Pachón in Chile is expected to see first light in 2019, with science operations beginning in 2022. This 8.4-metre (330-inch) monster is designed to capture detailed images of wide areas of the sky, completing the whole sweep in three days. It will repeat this over and over for ten years. It is hoped that comparisons across the scans will give a better picture of how dark matter (or modified gravity) plays out in practice across the universe.

Perhaps the most surprising possibility for finding out more about dark matter will also come from what is arguably the most exciting scientific development of the twenty-first century – the detection of gravitational waves. As we have seen (page 48), there are options for gravitational wave detection of some of the more obscure candidates for dark matter. Although primordial black holes are not a fashionable choice, they have not been entirely eliminated as a partial contributor to dark matter effects.

The LIGO gravitational wave observatory, based at Hanford, Washington and Livingston, Louisiana, began its third data run in its current configuration in 2019. It will then be upgraded. Similarly, other gravitational wave observatories around the world are being constructed and improved. We can expect a continuing flow of new gravitational wave data. In theory, these observatories could detect the merger of black holes smaller than the Sun. Such small black holes can't form in the normal fashion and would almost certainly be primordial. This doesn't mean they will be detected – but the potential is there to find them or perhaps eliminate them as a possibility.

Meanwhile there are other alternative approaches for dark matter detection being developed. A study published by the University of Surrey, Carnegie Mellon University and ETH Zürich in 2019 suggested that the dwarf galaxies that orbit larger galaxies like the Milky Way could tell us something about the behaviour of dark matter.

Specifically, the study looked at star formation in these dwarf galaxies. The formation process of stars can result in strong winds, moving gas and dust out of the centre of the galaxy. Although not directly affected by the wind or heat, any dark matter that has accumulated in the centre of the galaxy will be pulled by the gravitational attraction of large amounts of moving ordinary matter. This so-called dark matter heating (not a great name) should result in less dark matter at the centre of dwarf galaxies. The team observed sixteen dwarf galaxies and found that those which stopped forming stars many years ago had more central dark matter, seeming to support the theory. Because this is a new behaviour, when it is better quantified it could act as a test to distinguish between some dark matter theories.

Beware physicists seeking grants

At the time of writing, scientists at the CERN laboratory near Geneva are talking about the possibility of a next-generation collider, far larger than the current Large Hadron Collider. The proposal for the Future Circular Collider – at 100 kilometres long, nearly four times the size of the LHC – has an eye-watering £20 billion price tag and would have no guarantee of finding anything new. To justify such a project, however, the scientists involved have to come up with possible exciting new results, one of which is the suggestion that it could detect Higgs bosons in the process of decaying into dark matter particles.

This odd-sounding suggestion is based on what are

known as Higgs-portal models. These predict that dark matter should interact with ordinary matter by the exchange of a Higgs boson (much as electromagnetic interaction between ordinary matter particles involves the exchange of a photon). If that were the case, then it's expected that Higgs particles produced in a collider would sometimes decay into dark matter particles. This would make for a distinctive outcome as the decay products would be invisible.

This process has never been seen in the Higgs particles produced in the Large Hadron Collider, which means that theoreticians can rule out some candidates for dark matter particles. The justification, then, for the Future Circular Collider would be that it might observe this process – or eliminate still more dark matter candidates. But there's a problem here.

The whole concept is based on a hypothesis for which there is currently no evidence. We have no reason for assuming that the Higgs is involved in interactions between dark matter and ordinary matter. If that hypothesis is untrue, then the lack of Higgs invisible decays tells us nothing at all about the nature of dark matter particles. We're back to square one, having spent a lot of money.

The lesson seems to be that we should be wary of giving too much weight to such claims without careful examination of them.

Will we ever find dark matter?

A vast amount of effort has been put into finding direct evidence for dark matter, rather than the indirect hints provided by gravitational effects. As yet, the search has produced nothing concrete. As Vera Rubin ruefully pointed out in 2001, she had predicted back in 1980 that dark matter particles would be directly observed within ten years, yet they still hadn't been seen. Again, in the year 2000, the British Astronomer Royal Martin Rees made a similar ten-year prediction. His timescale has by now also been long passed.

In a book written late in his career, maverick astrophysicist Fred Hoyle unflatteringly portrayed his fellow scientists using a photograph of a running flock of geese, suggesting that once they have become hooked onto an idea, scientists will continue to follow it with no regard to new evidence. There's no doubt that Hoyle was bitter that his own steady state theory of the universe, developed in 1948 with Hermann Bondi and Thomas Gold, had been usurped in the 1960s by the big bang theory. Even Hoyle would have had to admit that the big bang was a better fit to evidence from the cosmic microwave background and the structure of very distant galaxies, seen early in their development because of the time light takes to reach us, than was the steady state model. But he also had a point.

The big bang theory had also failed to match observation several times and had to be patched up with various modifications such as inflation. Hoyle's point was that the steady state theory could also have been patched in this way to match observations. But by the time this possibility was explored, the cosmology community had gone into goose flock mode on the big bang and weren't interested. This is not dissimilar to the situation described by Thomas Kuhn in his flawed but highly influential book *The Structure of Scientific Revolutions* (1962).

Kuhn describes the progress of science as a process made up of occasional total changes of perspective, known as paradigm shifts (a term he didn't invent, but popularised), interspersed with periods of time when there is an accepted viewpoint. In the time prior to a paradigm shift, Kuhn suggested that there would be increasing evidence that made the status quo untenable, but as the current generation of scientists had so much invested in that status quo, they would prop it up as long as possible before the paradigm shift engaged.

The attitude to dark matter versus modified gravity seems to be in this state at the moment. In his blog Triton Station, Stacy McGaugh, professor of astrophysics at Case Western University, points out that there is a sociological issue in the way the 'dark matter or modified gravity' debate is being typified. He quotes a paper, 'A New Era in the Quest for Dark Matter' by Gianfranco Bertone and Tim Tait, in which the authors say: 'the only way [modified gravity] theories can be reconciled with observations is by effectively, and very precisely, mimicking the behavior of cold dark matter on cosmological scales.'

What the authors appear to be saying, McGaugh suggests, is that our current cosmological model, featuring the cosmological constant and cold dark matter (ΛCDM) is so successful that dark matter must exist. However, McGaugh believes they have the picture back to front, and a more accurate description is that the ΛCDM model is correct if and only if dark matter exists. McGaugh then playfully turns Bertone and Tait's assertion on its head as 'the only way ΛCDM can be reconciled with observations is by effectively, and very precisely, mimicking the behavior of MOND on galactic scales'.

That clash of viewpoints reveals a real problem. As we have seen (page 66), there are clear cases such as the galaxy NGC 1560 which fit modified gravity far better than dark matter. Some of these examples are significantly harder to explain if dark matter exists than counter-examples such as the Bullet Cluster are to explain using modified gravity. And though the deviation from dark matter's predictions is most obvious in the case of NGC 1560, in *the vast majority of galaxies*, dark matter is less effective at matching the observed rotation curve than is modified gravity. To make dark matter fit requires significant tweaking of extra parameters that are only required to make use of dark matter. This is anything but a one-off.

A few years ago, it would have been hard to present an argument against dark matter. Now, although the majority of astrophysicists and cosmologists still prefer dark matter to modified gravity, and expect that the missing particles will be discovered at some point, there is more of a balanced view. It will take time. As McGaugh points out: 'many scientists who are experts on dark matter don't know what MOND is really, or what it does and does not do successfully.' Sociological change is slow to come. But if more and more experiments fail to detect dark matter particles, there may come a tipping point where modified gravity takes over as the generally accepted picture.

As the old saying goes, absence of evidence is not the same of evidence of absence. Just because we can't detect dark matter particles doesn't mean that they aren't there. However, it would also be a mistake to think that science works by making truly logical deductions. This is hardly ever possible in practice, because deduction requires perfect knowledge. If I know, for example, that all stars work by nuclear fusion and I see an object in the sky that shines but isn't powered by nuclear fusion, I can deduce that it isn't a star. Instead of deduction, though, most science works by induction. Where deduction produces a definitive proof by definition, induction works from the available evidence and says what is most likely.

Take a simple example – I can predict that Halley's comet will return in 2061. This comet was one of the first bodies for which an astronomical prediction was made based on Newton's laws. So far, ever since 1705, when Edmond Halley predicted the comet's return in 1758, the comet has lived up to the inductive reasoning that its orbit will bring it back every 75–76 years.* However, I can't *prove* that it will come back as expected. It could be knocked off course by a collision. Other solar system bodies could undergo changes that influence Halley's orbit sufficiently to change its orbital period. All I can really say is that the comet has always fitted this timing for the observations we have made to date, and what we know of the solar system suggests that it will do it again.

Bearing in mind the way that science uses induction, it is perfectly feasible, then, if enough experiments that *should* be able to detect dark matter particles fail to discover them, to suggest that they probably don't exist. Something similar happened at the end of the nineteenth century, when American scientists Albert Michelson and Edward Morley showed that the ether, the imagined medium that light waves travelled through, appeared not to exist. Their experiment showed an absence of evidence for the ether. When followed up by repeated experiments, widely confirmed elsewhere, it was accepted that there was no such thing.

---

* Halley predicted 76 years, whereas the current accepted value is around 75.3 years.

At the moment, MOND remains the leading opponent to dark matter. As we have seen, there are alternative modified gravity theories, such as emergent gravity. As yet this does not appear to be as strong as MOND, but it has only been around since 2017 and it could still be that a variant of emergent gravity will prove a challenger to MOND. And, of course, we have the hybrid approach of superfluid dark matter, which some think combines the best of both worlds. The game is certainly not won yet between dark matter and modified gravity challengers.

With the exception of the suggestion of a mathematical hiccup in the calculations (see page 76), which as yet does not have wide support, it is accepted that something is happening in the universe that produces an important effect. For the moment that will be ascribed to dark matter with the proviso that it could be a result of modified gravity (a proviso that is often forgotten). Whatever the cause, there is something here with a significant effect that needs to be further investigated.

Even the basics are questioned

The problem cosmology and astrophysics always face is that by far the majority of the work is very indirect and involves measurements with large amounts of uncertainty. It might seem that there's enough of a problem finding dark matter – but in practice, astronomers are even struggling to find sufficient actual matter out there.

When studying nearby galaxies, where we would expect that the measurements would be most sound, recent research has come up with figures suggesting that some galaxies only appear to have about one-third of the normal matter that would be expected. Even the Milky Way is lacking around half its expected normal matter. This is still very much a puzzle, especially as the more indirect measurements from the cosmic microwave background match up well with the amount of matter predicted from direct observations.

One of the leading theories for the reason behind 'missing matter' is that the dark matter halo has managed to attract a lot of ordinary matter to itself, so that rather than lying in the visible part of the galaxy, large amounts of perfectly ordinary matter, emitting light too faintly to be observed, is also in the predicted spherical halo around the galaxy.

In 2018, astronomers used the European Space Agency's XMM-Newton X-ray observatory to hunt for the faint signature of this matter in a number of galaxies. The results, though, were disappointing. There did appear to be some ordinary matter in the haloes, but only about a quarter of the missing matter was accounted for.

The missing matter could be seen as an impetus to give more consideration to the idea that the calculation was wrong in the first place (see page 76), but for the moment it is merely inspiring further searches. The best hope seems to be for the missing matter to be scattered through the universe, not located in galaxies. This makes the matter extremely hard to detect, but the same XMM-Newton telescope produced results later in 2018 suggesting that there is indeed a significant amount of gas between galaxies (known as the warm-hot intergalactic medium) that would account for at least some and perhaps a significant amount of the missing matter. A more sensitive satellite observatory, Athena, due to be launched in 2028, will help fill in data on this phantom material.

The dark data keeps coming

Like dark matter, observations have continued to flow in on the parameters of dark energy. One of the most important currently under way is the Dark Energy Survey. This is a project using the Victor M. Blanco telescope – a 4-metre instrument at the Cerro Tololo Observatory in Chile, which is equipped with a red to infra-red camera that has a very wide field of view. Involving five years of observing, the survey started in 2013 and was completed in early 2019.

Reflecting the complexity of the process, the results from that first year were released in 2017, bringing in data from 26 million galaxies across a chunk of the southern sky and taking in views from relatively close back in time to 8 billion years ago. The outcome does not entirely match observations from the Planck satellite. This survey pushes the balance to there being slightly more dark energy than was thought, reflecting a dark energy/dark matter/normal matter ratio of 74:21:5.

Although – combined with Planck – the Dark Energy Survey gives us a best picture of the amount of dark energy, it emphasises the uncertainty that is still present in this kind of data. If we take the entire range from both sources, to have 95 per cent confidence in the outcome (a very low confidence level by physics standards), the percentage of the universe that is dark energy could be anywhere between 80 and 57 per cent.

At the extremes of the data from the two sources, there is a considerable variation between this survey and Planck. One significant difference between the two is that the Dark Energy Survey covers the relatively recent past, where Planck is looking back 13.4 billion years – so while other evidence (see page 125) seems to discount quintessence, this may bring the concept back into the frame. However, it's worth bearing in mind that only one of five years of data has been processed yet. More data could stress the difference ... or wipe it out entirely.

As we have already seen (page 131), the Large Synoptic Survey telescope is expected to be scanning the whole visible night sky from Chile repeatedly from 2022, which as well as attempting to better quantify the impact of dark matter will also be able to monitor the rate of expansion of the universe more accurately than ever before.

But what *is* dark energy?

Bearing in mind the vast range of theories that have been proposed to explain dark energy, some of these hypotheses are regularly being eliminated as new data becomes available. There are so many theories with very little evidence to support any particular variant that it would be tedious to list them all, but to give one example, 2018 saw the demise of the symmetron field.

The idea here was that there was another field stretching throughout spacetime, just as is the case with the electromagnetic field, the Higgs field which helps give particles their mass, and more. In this case, the field would be giving the universe a push to keep expanding faster. Researchers at TU Vienna hoped to measure the impact of hypothetical symmetron particles from this field on ultra-cold neutrons, but none was observed. The experiment doesn't entirely rule out symmetrons, but it does establish that they don't exist in the mass range necessary to make dark energy work

– which was the whole point of dreaming them up in the first place.

However, there still remains a wide range of theoretical possibilities with no particular favourite or good way of establishing what is actually happening. And new theories are being added all the time. For example, at the end of 2018, researchers at Uppsala University proposed a new string theory-based explanation for the source of dark energy.

As we have seen, the supersymmetric particles suggested by string theory have previously been proposed as candidates for dark matter. But these particles have never been observed – and for that matter, string theory, which once dominated the attempt to unify quantum physics and general relativity, is itself losing some of its gloss.

String theory sounds delightfully simple at first sight. Instead of particles it suggests that everything is made of one-dimensional strings, which vibrate in different ways, producing all the forces and particles we experience. However, dig in a little deeper and there are huge problems with string theory. It requires a universe that looks suspiciously different from the one we experience, with nine or ten spatial dimensions. There is no way of distinguishing between 10^{500} different potential outcomes of the theory. And it makes no useful predictions.

There's no doubt that string theory is attractive to a certain kind of theoretician who is enthusiastic about the 'beauty' of the mathematics, whether or not it actually conforms to our universe. It has been pointed out that string theory works better if the cosmological constant is negative, and as a result, many theorists spend their days happily working on string theory with a negative cosmological constant, even though the observed cosmological constant is positive. As the theory is effectively pure mathematics, this kind of theoretical physics does not have to be based on reality. It can still produce mathematically interesting results, but not ones that explain the universe we live in.

Bearing all this in mind, the researchers from Uppsala have proposed a model where our universe rides on the edge of a bubble expanding in one of the extra dimensions that string theory requires. In this model, the matter we experience is just the ends of strings which reach out into this extra dimension. That this should happen is compatible with string theory, and the expanding bubble would act as the source of dark energy.

Cynics might suggest that this is not much better than saying we don't know what dark energy is, and the model depends on a very simplified picture of the universe. But it's an example of the ways that theoreticians are continuing to search for mechanisms to explain the existence of dark energy.

Merging the two

As with dark matter, conventional dark energy itself isn't the only game in town, though here the challengers are more speculative. In late 2018 Jamie Farnes from the Oxford e-Research Centre in the Department of Engineering Science published a paper that he suggested not only explained dark energy, but did the double and accounted for dark matter as well.

Farnes proposed a new component of the universe, a fluid that had the property of negative mass. This gives it the effect of a negative gravitational force, meaning that it repels ordinary matter.* One of the problems with this approach is that by default, such a substance would get thinner as the universe expanded, reducing its effect rather than accelerating expansion as dark energy does. To cope with this, Farnes came up with what is described as a 'creation tensor' (tensors, as we have seen, are the generalised multidimensional mathematical structures also used in general relativity).

Some would consider the creation tensor to be something of a 'skyhook' – a made-up concept simply to make a model work – but it would not be the first time a kind of matter-creation mechanism has been proposed. (For example, in the steady state model of the universe, a new field was added to gravity, electromagnetism, etc. – a 'C-field' that caused matter to continuously be created.) As a by-product, Farnes' negative mass fluid would also explain some of the observations ascribed to dark matter.

At the time of writing there is little support for this theory. This is not entirely surprising, given both Farnes' discipline and the significant modification of our understanding of the universe that is employed here. Bear in mind how much resistance the Berkeley physicists received from astronomers when engaging in the search for supernovas to use as standard candles. Farnes is from an engineering department.

The other issue is the huge assumptions in play. Although both negative mass and matter-creation have been posited before, there has never been any evidence for them actually existing in the universe. Theoreticians come

---

* Such a material, should it ever be captured, would make it possible to build a perpetual motion machine, effectively generating work from nowhere.

up with abstruse models that don't fit with reality all the time. In this particular instance, negative mass may produce interesting results, but it brings with it some uncomfortable baggage.

Specifically, although the general theory of relativity does not rule out negative mass, it is clear about how such objects would behave. If we allow both positive and negative mass, then mass becomes a little like an inverse version of electrical charge. We are familiar with the idea that like electrical charges repel each other and opposite charges attract. In the case of mass, like masses attract each other but opposite masses should repel. So negative mass objects should be attracted to each other – but in the Farnes paper, they repel each other, which simply doesn't fit with general relativity.

It's very unlikely that this idea has any continuing merit, and it has been suggested that it produces problems more dramatic than the original problems it was thought up to solve. However, it is only by coming up with what can seem initially outrageous suggestions that may or may not prove valuable long-term that we can make progress. Ideas like this need to be treated with a lot of caution, rather than given the kind of misleading headline coverage they often receive in the press. But they are useful to the progress of science.

Dancing in the dark

It might seem that as this book draws to a close it has been a study of epic failure. We still don't know what dark matter is or even if it exists. We still don't know what dark energy is, and the cosmological constant is a factor of 10^{120} away from prediction. We are, appropriately, still in the dark.*

However, I see the current state of our understanding of dark matter and dark energy as a positive. Towards the end of the nineteenth century, physics student Max Planck, who was also a concert-level pianist, was told by his physics professor Philipp von Jolly that he ought to pursue music rather than physics, as all that was left for the physicist was to refine detail and add decimal places to observations. There was nothing original left to discover. Within a couple of decades, what are now the central tenets of physics – quantum theory (begun by Planck himself) and relativity – would be introduced and would totally transform what was previously known.

Some have been tempted to say that science really has achieved Jolly's 'near completion' now. Yet I would suggest that dark matter and dark energy demonstrate very effectively that there is far more yet to do. They are not alone in this. For example, we are still to unite quantum physics and the general theory of relativity. Although the Higgs boson has been found, its mass doesn't fit well with our standard model of particle physics, suggesting that either other new particles should exist (which stubbornly refuse to turn up) or that the standard model is fundamentally flawed. We have no idea how consciousness works or how either simple life or complex life can begin. Plenty of questions remain to be answered.

For me, the current level of ignorance is not a matter for depression, but delight. We have learned vast amounts in

---

* Some more recent calculations put the difference between observation and theory at 'only' 10^{50} or 10^{60} – but either way it's not exactly a trivial issue.

science in the last two hundred years, yet there is so much more still to discover. The universe would be boring if we knew everything – if there were no new intellectual frontiers to challenge us. The great, dark holes in our understanding of the universe, dark matter and dark energy, remain as stimulating as ever. Just as Sherlock Holmes was energised by taking on a new client, so scientists around the world can look to dark matter and dark energy as challenging mysteries worthy of their efforts.

We live in an age of science. Remarkably, around 90 per cent of the scientists who have ever lived are alive today. It's only right that there should be major challenges for them to face.

The dark game is afoot.

APPENDIX

Standard Model of Elementary Particles

Adapted from an image released by Fermilab, Office of Science,
United States Department of Energy, Particle Data Group

FURTHER READING

Chapter 1: Things ain't what they seem to be

Ignorance, Stuart Firestein (OUP, 2012) – explores the way that ignorance drives scientific discovery.

The Reality Frame, Brian Clegg (Icon Books, 2017) – explores relativity and our relationship to the universe.

Chapter 2: Exploring the universe

Astroquizzical, Jillian Scudder (Icon, 2018) – a good basic introduction to the universe.

Gravitational Waves, Brian Clegg (Icon, 2018) – entry in the Hot Science series covering the discovery of gravitational waves and how they might be used to find out more about the universe.

Chapter 3: The matter of missing matter

Cosmic Impact, Andrew May (Icon, 2019) – entry in the Hot Science series dealing with impacts of asteroids and comets on the Earth, including that associated with the extinction of the dinosaurs.

Dark Matter and the Dinosaurs, Lisa Randall (The Bodley Head, 2015) – a detailed exploration of the possible link between dark matter and the extinction of the dinosaurs from the theory's proponent.

Chapter 4: How big is the universe?

Astrophysics for People in a Hurry, Neil deGrasse Tyson (Norton, 2017) – a friendly approach to astronomy and cosmology.

Before the Big Bang, Brian Clegg (St Martin's Griffin, 2011) – explores the big bang and the theories of what might have come before.

The Beginning and the End of Everything, Paul Parsons (Michael O'Mara Books, 2018) – a good overview of current ideas on cosmology.

Chapter 5: Getting bigger faster

The 4-Percent Universe, Richard Panek (Oneworld Publications, 2012) – considerable detail on the individuals involved in the search for dark matter and dark energy, though already a little out of date.

The Cosmic Web, J. Richard Gott (Princeton University Press, 2016) – relatively technical but approachable overview of large-scale cosmology including dark energy.

Chapter 6: A continuing story

The news on dark matter and dark energy changes practically weekly – though often it's just more theories or more conflicting data. A good feel for the latest information can be obtained by searching the following websites for 'dark matter' and 'dark energy':

www.physicsworld.com
www.quantamagazine.org
www.sciencedaily.com

INDEX

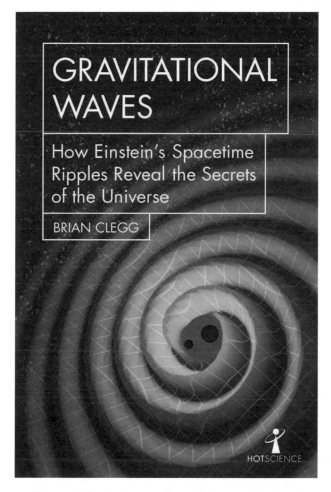

GRAVITATIONAL WAVES

How Einstein's Spacetime Ripples Reveal the Secrets of the Universe

BRIAN CLEGG

HOTSCIENCE

In 2015, after 50 years of searching, gravitational waves were detected for the first time. These ripples in the fabric of spacetime – predicted by Einstein and caused, in this case, by two black holes spiralling into each other – open up huge possibilities for the future of astronomy. Brian Clegg presents a compelling story of human technical endeavour and a new, powerful path to understand the workings of the universe.

ISBN 9781785783203 (paperback) / 9781785783210 (ebook)

CERN AND THE
HIGGS BOSON

The Global Quest for the
Building Blocks of Reality

JAMES GILLIES

HOTSCIENCE

The Higgs boson is the rock star of fundamental
particles, catapulting CERN, the laboratory where it
was found, into the global spotlight. But what is it,
why does it matter, and what exactly is CERN? James
Gillies tells the gripping story of one of the most
ambitious scientific undertakings of our time.

ISBN 9781785783920 (paperback) / 9781785783937 (ebook)